AF882648

AMERICAN REPRESENTATION IN OCCUPIED GERMANY

1920 - 1921

VOLUME II

COMPILED BY THE ASSISTANT CHIEF OF STAFF, G-2

AMERICAN FORCES IN GERMANY

AMERICAN REPRESENTATION IN OCCUPIED GERMANY 1920-1921

Vol. II

CORRECTIONS IN FOOTNOTES TO CONFORM TO MIMEOGRAPHED EDITION:

Page 23, fn 4 AMG, I, ch 1 and IV, App. No. 56 p. 381

" 68, fn 1 AMG, I, pp. 127-32

" 105, fn 1 AMG, I, p. 193

" 117, fn 1 AMG, I, p. 196

" 135, fn 1 AMG, III, p. 22
" " , " 2 Ibid., p. 32
" " , " 3 Ibid., p. 36
" " , " 4 Ibid., p. 40
" " , " 5 Ibid., p. 52

" 137, fn 1 AMG, III, p. 32

" 216, fn 1 For Inter-Allied Rhineland High Commission, see AR in OG, I, ch 2, pp. 145 ff
" " " 2 AMG, I, p. 50 and IV, p. 156

" 217, fn 1 AMG, IV, p. 203

" 219, fn 2 AMG, IV, p. 280

" 222, fn 1 AMG, I, p. 52 and IV, p. 215

" 295, fn 2 AMG, I, p. 92

" 297, fn 1 AMG, I, p. 101
" " , " 2 Ibid, p. 100

" 298, fn 2 AMG, I, pp. 98-100

" 360, fn 1 AMG, I, ch 3, pp. 39ff.

American Representation in Occupied Germany, 1920 - 1921.

Volume II.

TABLE OF CONTENTS.

AMERICAN REPRESENTATION IN OCCUPIED GERMANY.

1920 - 1921.

VOLUME II.

CHAPTERS.

CHAPTER NO. 3

THE OFFICE OF CIVIL AFFAIRS.

Transfer of Duties; Reorganization; Activaties - Executive Officer, Legal Department, Department of Fiscal Affairs, Public Utilities; Population of American Area; Appendices.

REFERENCE.

General . . . . A.M.G., Volume I, Chapters 4, 14, and 18.

The Interallied Rhineland High Commission..Volume I, Chapter 2.

THE OFFICE OF CIVIL AFFAIRS.

TRANSFER OF DUTIES.

On January 10, 1920, the Rhineland Convention went into effect throughout the entire territories occupied by the troops of the Allied and Associated Powers. Although the United States Senate had not ratified the Treaty of Versailles, with the consent of the Commanding General of the A.F.G. the Ordinances of the Inter-allied Rhineland High Commission were to be published as orders in the American area and to have full effect as such. The first Ordinances were published on January 10, 1920, in General Orders No. 2, which also revoked all orders previously published for the guidance of the civil population and which conflicted with the Ordinances.

Up to this time matters pertaining to the civil government in the American area had been taken care of by the Army through the Office of Civil Affairs. The entrance of the High Commission upon its duties involved the transfer of civil authority from the Army to the American Department of the High Commission. Such a transfer required that many difficulties due to the peculiar status of the A.F.G. be overcome and that there be as little confusion as possible in the minds of the German officials charged with the execution of the orders and regulations of the occupying authorities. Technically, the Commanding General of the American forces was still the supreme authority in the American zone, and although the Ordinances of the High Commission had been published as military orders therein, the American authorities were still responsible for their execution. A large amount of detailed work was placed upon the Office of Civil Affairs at this time, since such an important governmental change could not be accomplished over night. The responsibility for the proper functioning of the civil administration necessarily rested with the Army until suitable plans had been formulated for the transfer of that responsibility to the High Commission.

Under the Ordinances various duties were to be performed by the representatives of the High Commission located in the Kreise. These duties had theretofore been performed by the Officers in Charge of Civil Affairs in the Kreise, and it seemed a question whether the representatives of the High Commission could perform the duties assigned them by the Ordinances without creating in the minds of the German officials a loss of respect for the Officers in Charge of Civil Affairs. The following duties were performed by the Kreis Officers in Charge of Civil Affairs at that time (January 10, 1920):

Administration.

Supervision of prisons.
Investigation of appeals for clemency.
Issuing orders for deportation from American Area.
Investigation of complaints of civilians against the military.
Supervision of local circulation.
Supervision of civil administration.
Supervision of sanitation.
Zone Major duties - investigation and payment of claims.

Judicial

Provost Court duties, actually performed by an officer designated as Superior Provost Court and attached to the local Office of Civil Affairs;

Vagrancy Courts.

Surveillance of Civil Courts.

Economic

Surveillance of local food and fuel situation.
Supervision of industrial meetings.
Mediation between industries and the military forces.
Mediation (in emergencies) between employers and employees.
Surveillance of labor conditions and unemployment.
Preparation of weekly report.

Political

Supervision of political meetings.
Surveillance of local political tendencies.
Liaison between military forces and civil officials.

Outline of weekly report submitted by *Kreis* Officers in Charge of Civil Affairs:

I. POLITICAL.

(a) Feeling of the people toward carrying out the terms of the Treaty of Peace.
(b) Attitude of the people toward Ebert Government.
(c) Feeling of the people in regard to the withdrawal of the French from Frankfort.

II. FOOD.

(a) Condition of the crops in the *Kreis*.
(b) Changes and substitutions in the ration, and food shortages.
(c) Extent to which food laws were obeyed.
(d) Number of horses, cattle, sheep and swine handed over to the Allies under the Peace Treaty.

III. FUEL.

(a) The normal requirements of the *Kreis*.
(b) The average weekly amount reaching the *Kriese*.
(c) Comments on fuel situation.

IV. INDUSTRY AND TRADE.

(a) Opinion of merchants on market conditions. Stocks on hand.

(b) Relation of prices to the value of the mark.

(c) Report of any factory shut down because of lack of coal, and number of employees out of work as result.

(d) Strikes and labor disputes.

(e) Unemployment.

V. POPULATION.

(a) Attitude of the people toward their local officials.

(b) Biographical sketch of one Kreis offical per week.

By agreement between the American Representative with the High Commission and the Commanding General, dated January 17, 1920, it was decided that with few exceptions the duties assigned under the Ordinances to the representatives of the High Commission should actually be performed by them. Thereupon certain instructions were given to the Officers in Charge of Civil Affairs in the various Kreise. These instructions were briefly as follows:

German officials might be suspended by the High Commission in accordance with the Ordinances, the approval of the High Commission being requested should the Army desire to take such action against any official.

Representatives of the Army and the High Commission should make joint inspections of German prisons.

The proper German authorities were required to submit reports to both representatives (of Civil Affairs and of the High Commission), containing the names of persons entering the American zone or desiring to take up a residence therein.

The Army reserved the right to expel undesirable or dangerous persons from the American area, the High Commission ordinarily being requested to issue the order for deportation.

Notification of all meetings would be submitted to the Army representative as before and until such time as the High Commission should pass a suitable Ordinance covering the matter. In the case of political meetings, the representative of the High Commission would be at liberty to attend with the representative of Civil Affairs.

Permits to carry arms would still be issued by the Provost Marshal.

Requests from the German officials for the aid of a military force to establish order, when it could not be accomplished by the German police, would continue to be made to the Office of Civil Affairs in the *Kreis*.

Reports required from officials (concerning public order and conditions in the areas under their supervision) would be made to both representatives.

The Ordinances of the High Commission were to be published to the civil population under the direction of the Office of Civil Affairs.

These instructions adjusted the indefinite relations which had existed between the Officers in Charge of Civil Affairs and the representatives of the High Commission in the various *Kreise*, and misunderstandings by German officials as to the status of both authorities were thus avoided.

An examination of all military orders issued prior to January 10, 1920, was made at this time, with a view to the revocation or modification of those which conflicted with the Ordinances. On January 31, 1920 Civil Affairs Bulletin No. 58[1] was published. This bulletin revised or revoked previous orders and indicated the changes thus made; all in order that the civil administration in the American zone might conform to the Ordinances.

Until the beginning of June 1920 the duties of the Officers in Charge of Civil Affairs in the various *Kreise* remained practically the same, with the exception that by General Orders No. 19 (March 4, 1920), publishing Instruction No. 7 of the High Commission, the issuing of permits to carry arms was placed entirely in the hands of the High Commission (American Department.). In accordance with a memorandum from the Officer in Charge of

[1] Appendix I to this Chapter.

Civil Affairs, dated June 7, 1920[1], the duties of the various Officers in the Kreise were to be as follows:

(a) Administration

Supervision of prisons.
Investigation of appeals for clemency.
Issuing orders for deportation of vagrants.
Investigation of complaints of civilians against the military.
Supervision of civil administration.
Liaison between the military forces and the civil officials.
Supervision of sanitation.

(b) Judicial

Provost and Vagrancy Court duties.
Surveillance of the German Courts.

By this memorandum the entire supervision of the civil police in the American area was turned over to the Provost Marshal. The duties of Zone Majors were in the future to be under the Rents, Requisitions and Claims Service[2], but towards the end of the month the handling of claims was again placed under the Office of Civil Affairs. After the issue of the memorandum the American Department of the High Commission took over the supervision of economic and political matters in the area.

On October 21, 1920, the following order was issued:

1 Appendix 2 to this Chapter.

2 Volume II, Chapter 10.

HEADQUARTERS
AMERICAN FORCES IN GERMANY
Coblenz, Germany.

ORDERS
No. 192.

21 October, 1920.

FUNCTIONS OF CIVIL AFFAIRS IN THE AMERICAN ZONE TRANSFERRED TO THE INTERALLIED RHINELAND HIGH COMMISSION.

1. All duties heretofore performed by military officers in charge of Civil Affairs for the Kreise located within the American Zone, will, in the future, be performed by the representatives of the Interallied Rhineland High Commission located in the various Kreise.

2. Military Courts will continue to function as heretofore.

By Command of Major General ALLEN:

J. C. MONTGOMERY
Chief of Staff.

OFFICIAL:
S. G. TALBOTT
Adjutant General.

This order removed from the jurisdiction of the Officers of Civil Affairs in the Kreise all matters between the civil population and the military, with the exception of the trial of persons violating the orders of the Commanding General or the Ordinances of the High Commission.

There were, therefore, two phases in this transfer of civil authority during the period from January 10, 1920, until the latter part of October 1920: the first, from January 10 to June, when the Kreis Officers of Civil Affairs retained, for the most part, all of the authority vested in them previous to the coming into power of the Interallied Rhineland High Commission; the second, from June to October, during which time their duties, though still pertaining to civil affairs, were more largely of a military character. After this the American Department of the High Commission was entrusted with practically all the responsibility for the supervision of the civil government of the American area.

REORGANIZATION.

Upon the coming into existence of the High Commission the Office of Civil Affairs was reorganized as follows:

Designation	Function
Officer in Charge of Civil Affairs,	General Supervision and Policy.
Executive Officer,	Co-ordination.
Representative in office of Oberpraesident,	Liaison with German officials.
Department of Sanitation and Public Health,	Supervision of civilian health conditions.
Legal Department,	Legal decisions, Superior Provost and Vagrancy Courts and records of same, Clemency appeals.
Department of Fiscal Affairs,	Handling funds received from Court fines, Sale of enemy war material and confiscated property.
Historical Section,	Historical Records of Civil Affairs.
Economic Section,	Food, fuel and labor conditions; Industry.
Officers of Civil Affairs in Kreise: Kreise Adenau Ahrweiler Coblenz (Stadt and Land) Cochem Mayen Montabaur Neuwied	As previously shown.

By the first of June 1920 liaison with the German officials was maintained by the Department of Fiscal Affairs and the Historical and Economic Sections had been abandoned. In April the Provost Court, Kreis Coblenz, had been transferred to the Office of Civil Affairs, A.F.G. On June 7, 1920, the area was redivided into four Districts, each district comprising of two Kreise[1]. In October 1920 the representatives of the High Commission having taken over all their duties, the Officers in Charge of Civil Affairs in the Kreise were relieved. The Provost Court work in the American area, with the exception of Kreis Coblenz, was thereafter performed by Superior Provost Court No. 2, located in the town of Mayen. The court work in Kreis Coblenz was handled by Superior Provost Court No. 1, located in the Office of Civil Affairs, A.F.G. By the end of October 1920 the various departments, with the exception of the Department of Sanitation and Public Health, had been merged and placed directly under the supervision of the Officer in Charge of Civil Affairs, A.F.G., assisted by his Executive Officer. From October 1920 to the close of the year 1921, the organization of the Office of Civil Affairs remained as outlined above.

ACTIVITIES.

During the spring of 1920 the Officer in Charge of Civil Affairs was still largely responsible for the civil government of the American area. As the Interallied Rhineland High Commission

[1] Appendix 2 to this Chapter.

progressed with its organization, that body from time to time took over the duties previously entrusted to the Office of Civil Affairs. Due to the peculiar status of the A.F.G., many problems continued to arise; the policies concerning these were formulated by the Office of Civil Affairs. All Ordinances of the High Commission continued to be transmitted, as orders of the Commanding General, to the German civil officials by the Office of Civil Affairs. The office served as a link between the military and the civil population; complaints or inquiries received from civil sources were transmitted, when necessary, to the proper General Staff Division for consideration or action, and orders or regulations concerning civil matters were sent to the German officials for their guidance. Weekly conferences, attended by all officers on duty in Civil Affairs, were held in Coblenz. At various times visits were made to the British, Belgian and French areas.

Executive Officer.

Through the Executive Officer were submitted all matters for the action, or information of the Officer in Charge of Civil Affairs. The former was responsible for the proper co-ordination of the several departments and of the activities in the Kreise, for the dissemination of information and instructions, and for the receiving and rendering of reports and returns. Some of the routine duties in the early part of 1920 are contained in Appendix 3 to this Chapter. These activities greatly decreased as time went on. The matter of miscellaneous correspondence received from German civilians constituted a large portion of the duties of the Executive

Officer. Members of the occupying forces, mixing freely with the inhabitants, brought about many and varied situations which came to the hands of Civil Affairs for settlement. As it was the policy of the Commanding General that members of the command could not be subjected to civil suits in German courts, cases which would have involved Civil Suits, had they been permitted, were frequently adjusted by the Office of Civil Affairs.

Legal Department.

The Legal Department of the Office of Civil Affairs was charged with the general supervision of all Military Commissions and Provost Courts, together with the custody of all court records. It furnished opinions on legal questions and made recommendations upon pleas for clemency. Pleas for clemency, after recommendations had been made thereon, were approved or disapproved by the Officer in Charge of Civil Affairs. On February 21, 1921, a Clemency Board was appointed by the Officer in Charge of Civil Affairs; it consisted of one officer of Civil Affairs, the Representative of the High Commission for Kreis Coblenz and the Deputy Burgermeister of Coblenz. After that time all clemency appeals were acted upon by this Board.
Until the first of September, 1920 this department passed upon new German laws applying to the American area insofar as they might affect the occupying forces; after that time this became a function of the American Department of the High Commission. It may be said that no laws had been adversely reported upon by Civil Affairs. In September 1919 the Officers in Charge of Civil Affairs in the various _Kreise_ had been authorized to appoint Superior Provost Courts

and to review the cases tried by such courts. These Courts were set up in the various <u>Kreise</u> as the troops of the Third Army were returned to the United States, and they continued in operation after the coming into existence of the American Forces in Germany. By General Orders No. 76, October 6, 1920, Inferior Provost Courts were abolished and their jurisdiction transferred to the Superior Provost Courts.

A considerable number of cases brought before the Provost Courts were against persons charged with vagrancy.[1] Vagrancy (as defined by Civil Affairs Bulletin No. 34, October 15, 1919) consisted of persons whose usual place of abode being outside the American area were present therein without visible means of support, as well as women soliciting or having sexual intercourse with any person serving the United States Government or any associated Government. Upon trial and conviction, those persons domiciled without the area were deported in addition to any fine or sentence imposed. Persons under sixteen years of age, if the circumstances and nature of the case warranted, were turned over to the civil authorities to be placed in a reformatory, convent or industrial school, or to be returned to their parents or guardians. Civil Affairs Bulletin No. 58, January 31, 1920, limited the trials for vagrancy to those concerning vagrant women. If the circumstances warranted it, vagrant men were to be reported to the High Commission, with the request that they be deported from the occupied territories.

[1] Vagrancy Court duties were performed by the officers who sat as Superior Provost Courts. See A. M. G., Vol. I, Chapter 5.

Orders No. 37, April 1, 1920, again gave jurisdiction over vagrant men to the Provost Courts, but by General Orders No. 44, April 12, 1921, this jurisdiction was taken away, and as before vagrant men were reported to the High Commission for deportation.

Ordinance No. 6 of the High Commission provided that certain space on passenger trains in the occupied territories should be preserved for persons belonging to the forces of occupation and the High Commission, and made it a violation of the Ordinances for others to travel in spaces so reserved. Until October, 1921 unauthorized persons found in these compartments were required to remove themselves, but no disciplinary action was taken. As such cases became very frequent, it was decided to try all offenders. As security for their appearance before the Provost Court, they were required to deposit bail in an amount equivalent to $5.00 at the rate of exchange for that day. In case the accused person desired to continue on his journey, he could do so by forfeiting the bail deposited. Offenders against this Ordinance were not confined to await trial. This procedure resulted in a considerable decrease in the number of unauthorized civilians in reserved compartments.

From January 10, 1920, to December 31, 1921, a total of 8,476 cases were tried by the Superior Provost Courts; of this number 1,731 were acquittals and 6,746 convictions. The convictions were as follows:

Sale of prohibited alcoholic drinks...............	417
Unauthorized sale of light wine and beer..........	21

Unlawful possession of government property	437
Failure to obey lawful orders	134
Theft and similar offenses	340
Damaging military property	5
Obstruction	14
Assault	70
Insult	97
Harboring absentees	11
Unlawful use of uniform or badge	39
Offences against circulation regulations	241
Unlawful possession of fire-arms	142
Traveling in reserved military compartments......	30
Offenses against billeting regulations	51
Perjury	13
Overcharging or discriminating	67
Failure to salute	2
Violation of traffic regulations	156
Trespassing	2
Fraudulent claims	4
Miscellaneous	37
Vagrancy and prostitution	4416
TOTAL	6746

There were no trials by Military Commission during this period.

Department of Fiscal Affairs.

After January 10, 1920, this department was charged with the handling of payments of claims of civilians against the United States Government, and with accounting for fines received as the result of trials by Provost Courts and for funds received from the sale of confiscated liquor. The Act of Congress as published in General Orders No. 21, March 3, 1920,[1] provided for the payment of claims for damage to property done by American troops abroad, but in the case of claims for damages done by the troops in Germany, payment was made by the German Government after the claims had been acted upon and approved by the Rents, Requistions and Claims Service. All claims after being approved by this Service were sent to the Fiscal Department, Office of Civil Affairs, which forwarded them to the Oberpraesident of the Rhine Province for payment. The claims were sent by the Oberpraesident to the Burgermeister of the district in which the particular claimant lived; payments made by the latter were witnessed by the representative of the Office of Civil Affairs in the Kreis. Civil Affairs Bulletin No. 38 required that upon conviction for violation of the regulations concerning the sale of alcoholic liquor to the troops of occupation, the stock of liquor possessed by the person convicted would be seized and sold at public auction. The receipts from such sales were turned in to the Fiscal Department.

No sales of enemy war material were made subsequent to January 10, 1920.

The following amounts were received for the period January 10,

[1] Volume II, Chapter 10.

1920 to December 31, 1921[1]:

Fines, Provost Courts	847,688	Marks
Sales of Confiscated Liquor ..	6,791	Marks
Bail Forfeitures	5,315.50	Marks.
TOTAL	859,794.50	Marks.
Bail Forfeitures..	(186.50 Belgian francs) (70.00 French francs) (15.00 Dollars)	

Public Utilities[2].

During 1920 and 1921 the public utilities operated under the supervision of the Engineer Officer, A.F.G. Each week the Officers in Charge of Civil Affairs in the various Kreise (and later the representatives of the High Commission) submitted a report upon the public utilities within their respective Kreise. These reports were transmitted to the Engineer Officer for his information.[3]

POPULATION OF AMERICAN AREA.[4]

Kreis	1920	1921
Coblenz Stadt (City)	59,206	60,064
Coblenz Land	66,307	67,348
Cochem	41,053	41,160
Mayen	79,780	80,337

1 See Appendix 4 to this Chapter for total receipts from all sources since December, 1918.

2 See Volume II, Chapter 6.

3 See Appendix 6 to this Chapter, for a typical report.

4 A.M.G., Volume I, Chapter 1 and Appendix No. 56 Volume IV, Page 1463.

Adenau	24,837	24,871
Ahrweiler	49,053	49,945
Neuwied[1]	88,210	89,368
Unterwesterwald[2]	8,000	8,154
	416,446	421,247

APPENDICES.

1. Civil Affairs Bulletin No. 58, January 31, 1920.
2. Reorganization of Office of Civil Affairs, June 7, 1920.
3. Routine Duties of Civil Affairs, - 1920.
4. Total Receipts December 1918 to December 31, 1921.
5. Offenses against Civil Population.
6. A Report on Public Utilities.
7. Officer Personnel.

Appendix No. 1.

HEADQUARTERS
AMERICAN FORCES IN GERMANY
COBLENZ GERMANY

CIVIL AFFAIRS BULLETIN)
)
No. 58) 31 January, 1920.

REVISION OF MILITARY ORDERS.

In order to bring the orders of the military government as

[1] A small portion of Kreis Neuwied lay without the American area. The figures include only those living within the American area.

[2] The figures are approximate only. Only about one-sixth of Kreis Unterwesterwald was included in the American area.

far as practicable into harmony with the ordinances of the Inter-Allied Rhineland High Commission, which were published in General Orders No. 2, these headquarters, January 10, 1920, orders heretofore issued on the following subjects are revised or revoked as indicated. Where ordinances are referred to, those of the High Commission will be understood. It is the purpose of this order, taken in connection with General Orders No. 2, promulgating the ordinances of the High Commission, to bring to the attention of German citizens all existing orders of prime importance now in effect in the American Zone of Occupation.

1. ALCOHOL - DRUGS.

(a) Existing orders regarding the sale of alcohol or drugs to members of the American or Allied troops of occupation, being in conformity with the ordinances, will remain in effect.

(b) Only beer and light wines may be sold to persons in the military service (American or Allied) and to these only between the hours of 12:00 noon and 1:00 p.m., and between 5:00 p.m. and 9:30 p.m.

(c) All orders heretofore issued imposing restrictions upon the sale of alcohol either at wholesale or retail to persons other than members of the troops of occupation are revoked.

2. ARMS AND AMMUNITION.

(a) Orders heretofore issued requiring reports of sales of firearms and ammunition are revoked and the records and reports required therein will be made as provided in Ordinance No. 3, Part V.

(b) All arms heretofore turned in pursuant to orders of the military government may be returned to their owners upon complying with the ordinances.

(c) Permits to carry fire arms will hereafter be issued in accordance with the ordinances.

3. ASSEMBLIES AND MEETINGS.

The rules laid down in the ordinances will be carried out, except that notice of industrial meetings will also be given the same as required by the ordinances in the case of political meetings.

4. BILLETS.

Pursuant to Article 8 of the Rhineland Agreement, billeting will continue to be done in accordance with regulations issued from time to time by these headquarters. Orders heretofore issued requiring that no German nationals may reside in the city of Coblenz for more than twenty days without first applying to the "Staedtische Wohnungsamt, Rathaus," are continued in effect.

5. BLOCKADE - IMPORTS AND EXPORTS.

All restrictions heretofore issued from these headquarters regarding the export or import of goods, except arms, munitions and other material especially prepared for war, are removed.

6. CINEMATOGRAPHS - OBSCENE PICTURES, ETC.

The display of any obscene moving or still pictures in any cinematograph or any other public place of amusement is forbidden. The sale or gift or any attempt to sell or give away to members of the American or Allied Armies any indecent or obscene pictures, post cards, or literature of any kind is strictly forbidden. This order is supplementary to Ordinance No. 3, Article 13.

7. CIRCULATION.

Civilian circulation will be governed by the ordinances.

8. CLAIMS.

The settlement of claims for damages to property caused by American troops or military operations will be governed by rules issued from time to time by these headquarters. All civil officials are warned that it is the policy of the American Government not to permit the payment of any claims for damages alleged to have been committed by American troops until the same shall have been presented to and passed upon by the proper claims department of the Army.

9. CLEMENCY.

All appeals for clemency on the part of civilian prisoners sentenced by military courts will be presented to the Officer in Charge of Civil Affairs of the Kreis, who will investigate the same, indorse thereon his recommendations, and forward same to the Officer in Charge of Civil Affairs, American Forces in Germany, for the action of the Commanding General. If such an appeal for clemency is based upon physical disabilities it will be investigated by the nearest medical officer of the Army and the report of his examination will be forwarded with the appeal. Except under very unusual circumstances an appeal for clemency in the case of a civilian prisoner sentenced by a military court will not be considered in any single case more frequently than once a month.

10. COMPLAINTS AGAINST MILITARY PERSONS.

All complaints against members of the American Army must be reduced to writing and submitted to the Officer in Charge of Civil Affairs in the Kreis, who will make a thorough investigation of the facts and forward the same with his recommendations to the Officer in Charge of Civil Affairs, American Forces in Germany, for the information of the Commanding General.

11. COURTS - GERMAN.

(a) German civil courts will hereafter be permitted to exercise complete jurisdiction vested in them by German law or regulations without any supervision from the military authorities, subject only to such regulation as is provided for in the ordinances; provided, that until such time as the United States shall be at peace with Germany no member of the American Army, nor any person attached to or serving with the samo, shall be subject to the jurisdiction of any German court for any cause whatsoever.

(b) German criminal courts shall likewise be permitted to exercise their jurisdiction without military supervision, subject only to the provisions of the ordinances.

12. COURTS - MILITARY.

(a) The jurisdiction of Provost Courts shall remain unchanged.

(b) Vagrancy and Juvenile Courts: - Jurisdiction at present exercised by these courts will hereafter be limited to the trial of vagrant women as they are defined in orders heretofore issued from these headquarters; vagrant men shall be reported to the High Commission for deportation; juveniles shall be turned over to the German courts for trial.

13. ELECTIONS.

Any restrictions on elections heretofore imposed by orders issued by these headquarters are revoked.

14. FOOD LAWS.

All orders heretofore issued designed to assist in the enforcement of the German food laws, are revoked, except that for the present the military authorities will continue to punish as an offense against the military government any person who assaults or physically resists any official connected with the administration of the German food laws while the latter is acting in the performance of his official duty.

15. HUNTING.

All restrictions imposed by military orders upon Germans in the matter of hunting are revoked and this sport will be regulated hereafter according to the ordinances.

16. IDENTIFICATION.

(a) Identity cards (ausweise) will be provided and carried according to the ordinances. The attention of all German citizens is invited to the fact that in the first edition of the ordinances through typographical error in the German text it was not required that the "ausweise" have a photograph attached. This requirement must be observed.

(b) Orders heretofore issued requiring the heads of households to keep a list posted on the inside of the outer door of persons belonging to his household and residing in the building, are revoked.

17. MARRIAGE.

No person in the American military service is permitted to marry without the authority of the Commanding General, American Forces in Germany. Orders heretofore issued prohibiting German civil officials from performing marriages between members of the American Army and nationals of any country are continued in force.

18. MEDICAL TREATMENT.

Orders heretofore issued prohibiting German physicians, surgeons, apothecaries or any other persons not belonging to the Medical Department of the American Army from treating or giving medical advice in the case of venereal disease to any member of the American Army, or to supply them, directly or indirectly, drugs, medicines or medical or surgical apparatus for treatment of such disease, are continued in effect.

19. MILITARY DRILLS.

Military drills in public schools in occupied territory have heretofore been prohibited. Such prohibition will remain in effect until duly modified.

20. PHOTOGRAPHS.

Orders heretofore issued prohibiting the taking of photographs out of doors except by permission of the military authorities are revoked.

21. PIGEONS.

All orders regarding the owning, maintenance or use of carrier pigeons are revoked.

22. POLICE - GERMAN.

The organization of local police reserves heretofore authorized by these headquarters may continue until completed, if so desired by the German authorities, or until replaced by any other system authorized by the High Commission. The creation of any other police reserves in the American Zone under whatever name, without previous authorization of the High Commission, is strictly prohibited.

23. PRICES - DISCRIMINATION.

The attention of all merchants and shopkeepers of every sort and description is invited to the ordinances prohibiting discrimination in prices as against members of the Allied Forces or officials. Orders heretofore issued requiring the posting of price lists in restaurants under the seal of the German police and marking of articles for sale in shops are continued in effect. Nothing herein shall be construed as authorizing the levying of luxury, wine or other direct taxes on any member of the Allied Forces or of the High Commission.

24. PRISONERS OF WAR - DEMONSTRATIONS.

Orders heretofore issued prohibiting official or public reception of German prisoners of war will remain in effect. The return of such prisoners must not be made the occasion of official reception or any public assembly.

25. PROPERTY - UNLAWFUL POSSESSION OF.

Offenses against the property of the United States or any of the Allied Governments will be prosecuted under the ordinances. On account of the constant thefts of Government property in the American Zone the attention of all German nationals is invited to the fact that the possession of any such property is prima facie evidence of illegal acquisition and that the burden of proving title rests upon the possessor. Persons who have acquired such property by legal purchase should provide themselves with proper bills of sale for same.

26. PUBLIC UTILITIES.

Any person who wilfully damages, in a manner likely to prejudice the security of the troops, any public utility whatever shall be punished by a military court.

27. PUBLICATIONS.

The control of the press and of all publications shall be governed by the ordinances.

28. SALUTES.

Orders heretofore issued requiring all German officials in uniform to salute American and Allied officers are continued in effect.

29. SANITATION AND PUBLIC HEALTH.

(a) Instruction No. 4 of the High Commission requires German local authorities to report immediately every case of contagious or infectious disease both to the representative of the High Commission of the Kreis and to the military authority in the locality. The reports required by this instruction to be submitted to the military authorities will be submitted by the Kreisarzt to the Officer in Charge of Civil Affairs in each Kreis.

(b) The orders heretofore issued requiring barber shops to comply with certain sanitary regulations before being permitted to cater to officers and soldiers of the American Army are continued in effect.

30. STRIKES - LABOR DISPUTES.

All orders heretofore issued on the above subject are revoked and such matters will be handled hereafter under the ordinances.

31. TAXATION.

Article 9 of the Rhineland Agreement prescribed that "no German direct taxes or duties will be payable by the High Commission, the Allied and Associated armies or their personnel." Orders heretofore issued prohibiting the levying of wine or luxury taxes upon the troops of occupation are therefore continued in effect.

32. TRAFFIC REGULATIONS.

All orders heretofore issued for the control of civilian traffic, except in the Stadtkreis of Coblenz, whether street cars, automobile or animal drawn vehicles are revoked.

33. UNIFORMS - UNLAWFUL WEARING OF.

(a) Members of the German military forces on the active list entering the American occupied zone by authority will wear their uniforms at all times in said zone.

(b) The attention of all German nationals is invited to the ordinance prohibiting anyone from wearing the uniform or badge of the Allied Armies or of the High Commission or any colorable imitation thereof. Returning German prisoners of war must conform strictly to this ordinance. They will not be permitted upon the return to their homes to appear in uniform. Within five (5) days after reaching their homes their uniforms must be divested of all military insignia or distinctive marks, whether of German or Allied character.

(NOTE: Par. 58, Civil Affairs Bulletin is the last of this series).

By command of Major General ALLEN:

WM. W. HARTS,
Chief of Staff.

OFFICIAL:

S. G. TALBOTT
Adjutant General.

Appendix No. 2.

HEADQUARTERS
AMERICAN FORCES IN GERMANY
Office of Civil Affairs

Coblenz, Germany,
7 June, 1920.

MEMORANDUM FOR: All Officers in Charge of Civil Affairs.

1. By authority of the Commanding General, the following organization is created for the administration of Civil Affairs in the American Occupied Area:

DISTRICT NO. 1: (Including the office of the Officer in Charge of Civil Affairs, A.F.G., Kreis Coblenz and Kreis Cochem); Headquarters, Coblenz:

Lieut. Col. H. S. Grier, G.S., Officer in Charge of Civil Affairs.
Major M. C. Stayer, M.C., Chief Sanitary Inspector.
Major John D. White, F.A., Superior Provost Court and Vagrancy Court Officer for District No. 1.
Captain R. H. Dunlop, Inf., Executive Officer.
1st Lieut. P. C. Harper, F.A., Finance, Reports and Liaison.
2nd Lieut. Carl H. Hackert, C. of I., Interpreter.

DISTRICT NO. 2: (Kreis Montabaur and Kreis Neuwied); Headquarters, Montabaur.

Captain R. P. Hall, C.A.C., in charge of the district with station at Montabaur.
Captain John F. Goodman, Inf., in charge of Civil Affairs, Kreis Neuwied, with station at Neuwied. - Supervisor of the Superior Provost Court and Vagrancy Court of Andernach.
Major Frank B. Lammons, Inf., Superior Provost Court and Vagrancy Court Officer for District No. 2 and for the town of Andernach.

DISTRICT NO. 3: (Kreis of Mayen); Headquarters, Mayen.

Captain J. W. Barker, C.A.C., in charge of the district.
Major W. L. McSpedden, Inf., Superior Provost Court and Vagrancy Court Officer for District No. 3, except the town of Andernach.

DISTRICT NO. 4: (Kreis Ahrweiler and Kreis Adenau); Headquarters, Ahrweiler.

Captain K. McCatty, Cav., in charge of the district.
Captain Reinold Melberg, C.A.C., Superior Provost Court and Vagrancy Court Officer.

2. Beyond the court jurisdiction herein given and the necessary supervision incident thereto, the Officer in Charge of Civil Affairs at Neuwied will have no further connection with the civil administration of Kreis Mayen.

3. Under the re-organization the following are the duties to be performed by the Officers in Charge of Civil Affairs in the various districts and the personnel serving under them.

(a) ADMINISTRATION.
Supervision of prisons;
Investigation of appeals for clemency;
Issuing orders for deportation of vagrants;
Investigation of complaints of civilians against the military;
Supervision of civil administration;
Liaison between the military forces and the civil officials;
Supervision of sanitation.

(b) JUDICIAL.
Provost and Vagrancy Court duties;
Surveillance of the German courts.

4. By the direction of the Commanding General, the following duties that have been performed in the past by the Officers in Charge of Civil Affairs in the various Kreise, will now be taken over as follows:

By the Provost Marshal, A.F.G.: - The entire supervision of civil police.
By the R.R.&C. Service: - All Zone Major duties, including the investigation and payment of claims.
By the Kreis Representative, Interallied Rhineland High Commission: - (a) Economic.
Surveillance of local food and fuel situation;
Supervision of industrial meetings;
Mediation between industries and the military forces;
Mediation, in emergencies, between employers and employees;
Surveillance of labor conditions and unemployment;
Preparation of the weekly report on food, fuel, industry and trade, and population.
(b) Political.
Supervision of political meetings and preparation of that part of weekly report devoted to politics;
Surveillance of local political tendencies.

5. All officers serving under the Officer in Charge of Civil Affairs are enjoined to cooperate in every way possible with the

Kreis representatives. It is to be understood that the Kreis Representative is not under the orders of the Officer in Charge of Civil Affairs of the district. The main idea to be kept in mind is that this arrangement is one of a partnership in which a specific division of duties has been made as indicated herein. Following out this principle, officers in charge of districts and the personnel serving under them will arrange to allow the Kreis Representative, in the discharge of his duties, to have a proportionate use of such transportation as may be in their hands.

6. Channels of communication for all officers serving under the Officer in Charge of Civil Affairs will continue as heretofore.

7. The foregoing assignment of personnel and the duties herein enjoined will become effective on June 10, 1921.

H. S. GRIER,
Lt. Col., G. S.,
Officer in Charge of
Civil Affairs.

Appendix No. 3.

ROUTINE DUTIES.

OFFICE OF CIVIL AFFAIRS -- 1920.

I. Reports from Civil Sources.

(a) Landrat, Landkreis, Coblenz.

Weekly strength report and report of changes of German Police for transmission to Superivisor of German Police.

Monthly strength report of German Police for transmission to Supervisor of German Police.

Weekly report of arrest and disposition of German civilians for information of Provost Court Officer.

Weekly report on police matters of Landkreis Coblenz.

Report of complaints of civilians against military.

Weekly report on food situation in Landkreis.

All matters arising in the Landkreis relative to hunting, arms permits, return of hunting weapons, etc.

Weekly report of progress made in the organization of Police reserves.

(b) Oberburgermeister, Coblenz.

Weekly report on food situation.

Report of complaints of civilians against military.

Weekly report of fuel received in Stadt and Land-Kreis Coblenz.

City Treasury, receipts for billet money.

Police Headquarters, Coblenz. Weekly report of arrest and disposition of civilians for information of Provost Court.

Kreis Arzt, Coblenz. Weekly report on sanitation.

(c) Miscellaneous.

Amtsgericht, Coblenz. Weekly report of cases tried for information of Provost Court Officer.

Landgericht, Coblenz. Weekly report of cases tried for information of Provost Court Officer.

Weekly report of sale of hunting arms and ammunition.

Daily report of diseased women convicted by Vagrancy and Juvenile Court, from German physicians.

Weekly report from Metternich, Weissenthurm and Engers Gas Works of fuel and gas situation for transmission to Chief Engineer, Public Utilities.

II. Reports from Military Sources.

Military Prison, daily strength report.

Supervisor of German Police, daily report of arrests, etc., in City of Coblenz.

Supervisor of German Police, weekly report on German Police of Coblenz.

Inferior Provost Courts, weekly report of trials.

Chief Sanitary Officer, daily report of diseased women convicted by Vagrancy and Juvenile Court.

Chief Sanitary Officer, report of transfers of women from hospital to jail.

III. Reports, Orders and Routine Correspondence.

Weekly report of Civil Affairs.

Daily disposition sheets of Superior Provost Court.

" " " for vagrants.

Disposition sheets for juveniles tried by Juvenile Court.

Special Orders - granting clemency.

Deportation Orders.

Denial of clemency - by letter.

Marriage licences between Americans and foreigners.

Miscellaneous correspondence transmitted to the Chiefs of Staff Sections concerned on every subject from permits to burn extra lights to the sale of rubber boots.

IV. Requisition of Supplies.
Management of office force, etc.

Appendix No. 4.

TOTAL FUNDS HANDLED BY THE OFFICE OF CIVIL AFFAIRS FROM BEGINNING OF OCCUPATION, 1918, TO DECEMBER 31, 1921.

Total fines collected up to December 31, 1921 . . . . . 2,694,798.37 Mks.

Receipts from forfeiture of bail....185.60 Belgian Francs
70.00 French Francs
15.00 Dollars ...... 5,315.50 Mks.

Miscellaneous receipts . . . . 1,200.00 Mks.

Receipts from sale of confiscated liquor . . . . . . 674,473.00 Mks.

Funds deposited with the Disbursing Quartermaster, acc ing from the sale of enemy war material sold in violation of the terms of the Armistice . . . . 6,416,489.52 Mks.

AGGREGATE RECEIPTS ------		9,792,276.39 Mks.
	Belgian Francs	185.60
	French Francs	70.00
	Dollars	15.00

Appendix No. 5.

TABULATION OF OFFENSES COMMITTED BY AMERICAN TROOPS AGAINST THE CIVIL POPULATION, SHOWING ACTION TAKEN AGAINST OFFENDERS.

	Offender not identified	Complainant at fault	No action taken	Acquitted	Convicted	Total Cases
Homicide	1	1	1	3	5	11
Rape	3			3	2	8
Attempted Rape	3			1	6	10
Misconduct to Women			1	1	1	3
Sodomy				3	6	9
Robbery	13		5	5	15	38
Burglary	9		2	2	4	17
Larceny	16		6	18	61	101
Assault with Deadly Weapon	9		4	13	33	59
Assault	118	8	26	64	213	429
Accident	18	49	12	18	16	113
Miscellaneous	62	4	24	15	78	183
Total Cases	252	62	81	146	440	981

\- - - - -

Appendix No. 6.

WEEKLY REPORT ON PUBLIC UTILITIES
FORWARDED TO CHIEF ENGINEERS, A. F. G.

For Kreis Montabaur.
For a week in August, 1920.

Hoehr Gas Works.

Coal on hand	29.4 tons
Coal consumed during week	13. "
Coal received during week	36. "

Production . . . . 28.07 cu. meters.

Number of connections from plant
Private: 276
Industrial: 30 . . 306
Number of inhabitants supplied with gas 2950

Baumbach Electric Works.

Coal on hand	27.4 tons
Coal consumed during week	2.9 "
Coal received during week	0. "

Population of area supplied	Baumbach	1030
	Ransbach	1802
Consumers	Baumbach	205
	Ransbach	360
Industrial	Baumbach	40
	Ransbach	31

Number of kw. hrs. generated 2830.5
Peak load . . . . 65. kw. per hour
Direct current, 3 connections. . Light 110 and 220 volts
. . Power 220 volts.

Distribution: Private 300 kw. hrs.
Industrial 2445 kw. hrs.

Montabaur Electric Works.

Coal on hand	34½ tons
Coal consumed during week	14½ "
Coal received during week	30 3/4 "

Population of area supplies . . . 4350
Consumers . . . . 823 (Private 794, Industrial 29)

Number of kw. hrs. generated . . . 4800
Peak load . . . . 100 kw. per hour.
Direct current, 3 connections. . . Light -- 110 volts
Power -- 220 volts.

- - - - - - -

Appendix No. 7.

OFFICERS ON DUTY WITH CIVIL AFFAIRS

between

JANUARY 1, 1920 and DECEMBER 31, 1921.

NAME	RANK	DUTIES
Hunt, I. L.	Colonel, Inf.	Officer in Charge of Civil Affairs.
Stayer, M. C.	Colonel, M.C.	Chief Sanitary Officer.
Stone, David L.	Colonel G.S.	Officer in Charge of Civil Affairs.
Goff, Guy D.	Colonel, J.A.G.D.	Legal Department.
Bensel, Walter	Lt. Col. M.C.	Chief Sanitary Officer.
Grier, H.S.	Lt. Col. G.S.	Officer in Charge of Civil Affairs.
Rall, R. P.	Lt. Col. C.A.C.	Acting Officer in Charge of Civil Affairs.
Goodrich, P.M.	Lt. Col. Inf.	Superior Provost Court, District 1.
Evans, Vernon	Major, Inf.	Executive Officer.
Smith, Truman	Major, Inf.	Liaison.
White, J. D.	Major, Inf.	Superior Provost Court, Kreis Coblenz.
McCatty; K.	Major, F.A.	Off. in Charge of Civil Affairs, Kr. Ahrweiler.
McSpedden, W.L.	Major, Inf.	Superior Provost Court, Kreis Montabaur.
Fiske, William	Major, Inf.	Off. in Charge Civil Affairs, Kreis Cochem.
Lammons, F. B.	Major, Inf.	Superior Provost Court, Kreis Neuwied.
Bissel, H. B.	Major, Inf.	Off. in Charge Civil Affairs, Kreis Adenau.
Royse, F. E.	Major, Inf.	Superior Provost Court, Kreis Cochem.
Lockwood, B.C.	Major, Inf.	Superior Provost Court, Kreis Montabaur.
Goodman, J.F.	Major, Inf.	Off. in Charge Civil Affairs, Kreis Neuwied.
Pinkerton, H.M.	Major, Inf.	Superior Provost Court, Kreis Coblenz.
Barker, J.W.	Major, C.A.C.	Off. in Charge Civil Affairs, Kreis Mayen, and Superior Provost Court, District 2.
Dunlop, R. H.	Major, Inf.	Officer in Charge of Civil Affairs, Kreis Coblenz, and Executive Officer.
Melberg, Reinold	Major, C.A.C.	Executive Officer.
Flynn, T. J.	Major, M.C.	Chief Sanitary Officer.
Downs, S.D. Jr.	Major, F. A.	Executive Officer.
Crea, H.B.	Major, Inf.	Superior Provost Court, District 2, and Executive Officer.

NAME	RANK	DUTIES
Barber, T.H.	Captain, Inf.	Fiscal Affairs.
Pinkerton, W.P.	Captain, Inf.	Superior Provost Court, Kreis Mayen.
Tower, W. L.	Captain, San. C.	Dept. of Sanitation and Public Health.
Banham, C.G.	Captain, C.A.C.	Superior Provost Court, Kreis Montabaur.
Hopps, Edwin S.	Captain, Inf.	Superior Provost Court, City of Andernach.
Page, R.C.M.	Captain, Inf.	Officer in Charge Civil Affairs, Andernach.
Harper, P.C.	Captain, Inf.	Fiscal Affairs and Executive Officer.
Bente, C.F.	1st Lt. Inf.	Executive Officer.
Bird, Clarence	1st Lt. C. of I.	Translations.
Bevans, S.M.	1st Lt. F.A.	Executive Officer and Superior Provost Court.
Hackert, C.H.	2nd Lt. C. of I.	Translations.

\- - - - - - -

CHAPTER NO. 4

RELATIONS WITH THE CIVIL POPULATION.

Fraternization; Marriage; Illegitimate Children; Offenses by Soldiers; Appendices.

REFERENCE.

General . . . . A.M.G., Volume I, Chapter 11.

RELATIONS WITH THE CIVIL POPULATION.[1]

FRATERNIZATION.

After the revocation in September 1919 of the "anti-fraternization order," the American troops and the civil population settled down to what might well have been called normal relations. Between the soldiers and the larger part of the civil population the relations were friendly. Occasionally, there would occur a short period during which relations would have become slightly strained.[1] But on the whole it seems doubtful if the general relations between the troops and the population of an occupied territory have ever been more friendly than were those that prevailed in the American area during 1920 and 1921. The term is used advisedly, for after a year's experience with the American soldier the attitude of the average German was nothing short of friendly. In fact it might almost be said that the relations were quite as good as those that exist in the United States between the Regular soldiers and the neighboring civilian inhabitants. There was practically an entire absence of social relations between the officers and the Germans, but the contrary was true of the soldiers insofar as certain classes of Germans were concerned. Their relatively high pay in the depreciated German currency and their free-handedness and generosity caused the American soldiers to have a peculiar attraction for the German girls and to be welcome visitors in many respectable homes. The well-behaved and self-respecting American soldier found himself a respected member of the community, and he was not slow to take advantage of a welcome that except in time

1 Appendix 2 to this Chapter.

of war his own countrymen have never extended him. At the same time he never forgot that he was the representative of a victorious people, he continued to preserve a tactful attitude of superiority, and in every way he showed himself worthy of the country from which he came. On the other hand, as was inevitable, there was a small number of soldiers whose conduct served to throw discredit upon the great mass of good soldiers who formed the rank and file of the A.F.G. Frequently under the influence of strong drink and consorting with a low type of prostitute, their actions, combined with the conduct of certain German hoodlums, were usually the cause of the unfriendly feelings that occasionally were aroused in the minds of small sections of the civil population.[1]

The departure of a train filled with soldiers bound for the United States furnished evidence of the friendly relations mentioned above. The sight of the throngs of Germans gathered about the train, of the sorrowful and in some cases tear-streaked countenances, and the shouted farewells made it difficult to realize that those leaving were soldiers of an army of occupation or that the crowds were composed of inhabitants of an occupied area. One could but reflect that the departing soldiers would probably meet with no such cordiality upon their arrival in their own country.

MARRIAGE.

Under all the circumstances it was but natural that many soldiers should have desired to marry German girls. Those who desired marriage were required to secure the written permission of Headquarters. The policies governing marriages of soldiers

[1] Appendix 2 to this Chapter.

were formulated by G-1 (the Personnel Division of the General Staff), and the action, in accordance with those policies, upon applications was a function of the Adjutant, A.F.G. After securing the written approval of his application, the soldier secured from the Office of Civil Affairs a letter to the German authorities stating that there was no objection to his marriage. Unless this letter were presented by the soldier, the civil ceremony required by German law could not be performed. After December 1920 the securing of this letter from Civil Affairs was not required; in its stead the approved application was submitted to the German officials. In the case of an officer or field clerk, the Office of Civil Affairs requested the competant German officials to waive the usual formalities and to perform the civil ceremony without delay.[1]

As soon as the ban on fraternization had been lifted, many soldiers requested permission to marry. There were five main factors that had to be taken into consideration before a governing policy could be formulated --

> The shortage of billets that could be used for the families of enlisted men.
> The limited accommodations for such families on Army transports.
> Pregnancy before marriage.
> The necessity of preventing soldiers from marrying women of ill repute.
> Ability of the soldier to provide living expenses for his wife, both in Germany and while en route to a new station in the United States.

The presence of the High Commission, with its very large personnel, had served still further to complicate the already

[1] Compliance with the German laws requiring a civil as well as a religious ceremony was necessary, in order that the validity of the marriage might never be questioned.

difficult billeting situation in Coblenz. There was a great congestion in the billeting accommodations there, while in the other towns garrisoned by American troops there were very few billets available for married enlisted men. It became necessary, therefore, to return to the United States as many married soldiers as practicable.

The second class accommodations on the transports that plied between Antwarp and the United States were limited. This fact caused a limit to be put upon the number of marriage applications that could be approved within a certain period of time.

A large proportion of the marriage applications stated that the girl was pregnant and that early marriage was necessary. In not a few of those cases it was highly probable that the pregnancy had been brought about deliberately in order that the necessary permission to marry might be secured. Again it had to be borne in mind that in Germany sexual intercourse during betrothal is not unusual in certain classes, and that if there is a sincere intention to marry the practice is not viewed in the same light as in the United States.

After the restrictions upon civilian circulation had been relaxed, professional prostitutes flocked to Coblenz. Under their influence - or sometimes, perhaps, unable to discriminate - soldiers occasionally requested permission to marry women of this class. Then, too, it became apparent that some loose women, attracted by the comparative affluence of the soldiers and cognizant of American ideas regarding responsibility for the pregnancy of an unmarried woman, had choosen pregnancy as a means of inducing or forcing marriage.

While the soldier, due to the depreciated state of the German currency and to the lower standards and cost of living to which his betrothed was accustomed, might be well able to support a wife in Germany, the problem of existence in the United States was another matter. The Government would transport his wife across the Atlantic,[1] but he had to defray all other traveling and living expenses, before sailing as well as after arrival in the United States. Thus the question of sufficient funds could not be ignored, especially as information from the Red Cross in New York stated that some married soldiers had arrived in the United States without funds and their wives had become objects of charity.

Another factor - and not the least important - was than in the large majority of cases marriage lowers the efficiency of the enlisted man. And it has been the common experience of military men that the larger the proportion of married soldiers in a military organization, the lower its efficiency.

Under the original policy an enlisted man of Grade 3 (Staff Sergeant) or above was permitted to marry, provided investigation showed that the woman was of good character and that both parties were free from communicable disease. Unless exceptional circumstances were shown, marriage applications of soldiers below Grade 3 were disapproved. The reasons for the adoption of this policy were shown in the following letter from the Commanding General to the War Department, under date of April 30, 1920:

> It has developed that soldiers as well as young German

[1] Railroad transportation was furnished only to the dependents of soldiers above Grade 4.

women are writing to the War Department and to members of Congress seeking authority to get married, especially where pregnancy is involved. In a recent case the girl stated in her communication that she had allowed herself to become pregnant that her lover might marry her. It developed many months ago that should premission be given in such cases we would have numerous married men in these forces, greatly detrimental to military efficiency. Having established a policy of allowing only soldiers above grade nineteen (now grade 4), I have refused authority to all others below that grade regardless of a state of pregnancy.

In spite of what a first thought, or a casual consideration of this matter would dictate, I am sure that the policy pursued is sound from the point of view of military efficiency. The numerous excess of goodlooking marriageable girls in this Zone over corresponding German males accentuated the necessity of adopting the policy cited to avoid having a partially Germanized command.

On July 12, 1920, the letter was returned by the War Department, with the following endorsement:

> The policy announced by you of permitting only soldiers above grade nineteen to marry German women is disapproved.

This decision of the War Department made necessary a change in the existing policy. After careful consideration of all factors affecting the question, a new policy was adopted and published to the command. The German authorities were informed as to the new policy and were requested to transmit the information to such Germans as it might interest. The gist of the policy was as follows:

> Marriage applications from enlisted men, regardless of grade, would be acted upon in accordance with the merits of each case.
>
> No application would be approved if investigation showed the women's character to be questionable. It must be shown that she would be desirable as the wife of an American soldier and as a citizen of the United States.
>
> The marriage of an enlisted man would not be authorized until he had deposited $100.00 with his organization commander. This amount was to defray the expenses of the wife from Coblenz to the soldier's new station in the United States. The remainder of these funds was to be delivered to the soldier upon his arrival in New York.

No married soldier below Grade 3, except those whose retention might be especially authorized, would be allowed to remain in the A.F.G.

Any soldier who should marry without authority would forfeit all rights to a military certificate[1] and to ocean transportation for his family.

During any one month no greater number of applications from soldiers below Grade 3 than the accommodations for their wives on the next month's transport warranted, would be approved.

The soldier's character must be "Very Good" or "Excellent."

Despite the policy of returning to the United States all married soldiers below Grade 3 and of limiting the number of approved applications in accordance with transport accommodations, marriages rapidly grew more numerous. At one time the number of soldiers' dependents was so large that, with the approval of the War Department, the troop space in one transport was converted into second-class accommodations for their use. This was but a temporary relief, as the authority for the conversion applied to but one transport. However, as long as individual soldiers rather than units were being returned to the United States, it was possible to limit the number of married men on each transport in accordance with the second-class accommodations thereon. The return of troop units, which was to begin early in 1922, would make it necessary to return their personnel intact, regardless of the number of married men. It was clear that some soldiers would have to sail without their wives, but it was planned to send their dependents to the United States as quickly as the transport accommodations would permit. After the receipt of orders prescribing the reduction of the A.F.G., all enlisted men who requested permission to marry were informed that no transport accommodations were available and that none could be promised. It seemed probable that unless troop space were again used for soldiers' dependents, it would be necessary for many enlisted men to transport their families by commercial liners or else to leave them in Europe.

The American Headquarters did not actually dictate whether or not a soldier should marry. The approval of an application meant simply that the dependents would be furnished ocean transportation if practicable and that, if such transportation should be furnished, the wife or other dependent would be given a Military Certificate. The Military Certificate was a paper, signed by the Adjutant, A.F.G., which stated that the person was a bona-fide dependent of an American soldier; the paper was recognized by the American immigration authorities as having the same value as a passport. The Military Certificate was a make-shift, made necessary by the fact that the technical state of war between the United States and Germany did not end until the ratification of a treaty of peace late in 1921. During that technical state of war, by a decision of the State Department German women could not become American citizens by virtue of marriage to American soldiers, and hence could not be given American passports. The Military Certificate continued to be issued after the ratification of the Treaty of Berlin, but it was planned to discontinue its issue in the spring of 1922.

Immediately after his marriage to a German woman, the soldier reported to the Adjutant, A.F.G., who prepared the certificate. Upon the departure of the soldier and his wife for the United States, the certificate was sent to the Quartermaster of the transport. Upon arrival in the United States, the latter delivered the certificate to the immigration authorities. It was valid only when the dependent was a passenger on an Army transport, and was never delivered to the dependent in person.

In general the policy with regard to the marriage of enlisted men worked satisfactorily, and it was probably the best practicable solution of the problem. Inasmuch as the great reduction of the A.F.G., which was to take place in 1922, had caused each new application to be advised that the transportation of his dependents could not be promised, it was expected that there would be a corresponding decrease in the number of marriages.

The following statistics are of interest:

Number of applications received for marriage from October 1, 1919 to January 1, 1922 . .	1527		
Number approved		767	
Number disapproved.		760	
		1527	
Number of military certificates issued from April 1, 1920 to January 1, 1922	595		
Number of enlisted men's dependents returned to the United States from 1 December, 1919 to 1 January, 1922	Women	Children	
	782	267	

Nearly 90% of these marriages were to German women. About 6% of the brides were French, and the remainder included girls of American, British, Belgian, Polish, Luxemburg and Swedish birth.

ILLEGITIMATE CHILDREN

The Office of Civil Affairs received many appeals from German women claiming that they had given birth to children whose fathers were American soldiers and requesting that the fathers be required to contribute to the support of the children. If the soldier was still in the service, the matter was referred to his immediate Commanding Officer for investigation. Should the soldier admit parentage, he was urged to assist in the support of the child by contributing monthly a sum equivalent to the amount required by the German law in such cases. If the soldier denied parentage or declined to contribute toward the support of the child, the mother was advised that legal action before a United States Court was her only recourse.[1] In a few of these cases the soldier secured permission to marry the woman, while in about 25% the soldiers agreed to make monthly contributions toward the support of the child. In many cases investigation showed that the claims were without foundation - women of ill repute often made such claimsin the hope of obtaining money for themselves alone.

If it were found that the father had been discharged from the Army, the woman was informed accordingly and was furnished with the address given by him at the time of his discharge. She was advised that she could take up the matter directly with the alleged father, and that if appeals to him should not accomplish the desired results, she should institute legal proceedings if such were practicable.

OFFENSES BY SOLDIERS

It was required by orders that commanding officers should investigate thoroughly all complaints alleging offenses by members of their commands against the persons or property of Germans. When an investigation demonstrated the guilt of a soldier, suitable disciplinary action was to be taken. Reports of all

[1] In accordance with the policy of the Commanding General, civil suits against American soldiers could not be brought before German courts.

such cases were made to the Office of Civil Affairs, where they were recorded.

In May 1921 the Judge Advocate was directed to make to the Office of Civil Affairs a monthly report, containing a list of all officers and soldiers tried during the month for misdemeanors or crimes against Germans and showing the result of each trial. It was also required that the commanders of regiments or other separate units should make similar reports regarding all cases brought to their official attention, showing that the soldier had been awarded company punishment or that other action had been taken. A record of the final action in all cases of alleged offenses against Germans was kept in the Office of Civil Affairs. That office made to the American Department of the High Commission a monthly report of all offenses upon which final action had been taken; in each case where there had been a conviction the punishment was shown in the report. The American Department then furnished similar information to the German authorities, thus demonstrating the endeavors of the Americans to punish soldiers who had been found guilty of offenses against Germans.

In Appendix 5 to chapter 3 is given a tabulation of the crimes and misdemeanors reported during 1920 and 1921.[1] Where the offenders were not identified it is highly probable that a number were not American soldiers. Many of the offenses were committed under cover of darkness, and the Germans displayed a marked tendency in case of doubt to allege that "some soldier" had been at fault.

[1] Under the heading "No action taken" are shown those cases the investigation of which showed that disciplinary action was not indicated.

In several cases it was found that a deliberate "frame-up" had been attempted, while it was well known that many Germans went about at night in American uniforms. As a precaution against fraudulent claims, and at the same time to insure the making of proper records in cases of alleged injury by American soldiers, General Orders No. 77 were issued on October 7, 1920 --

PROCEDURE IN CASES OF INJURY OR ACCIDENT TO PERSONS IN AMERICAN OCCUPIED AREA.

1. In all cases of injury or accident to persons in the American Occupied Area resulting from the actions of persons belonging to the American Forces in Germany, a Medical Corps Officer of the United States Army will examine, with the least possible delay, the nature of the injuries. He will submit through military channels a full report as to the conditions of the injured party and the circumstances attendant upon the infliction of the injury or accident.

2. The Officer in Charge of Civil Affairs, will take the necessary steps to make known these instructions to the proper German officials and inform them that an American Medical Corps Officer should be summoned from the nearest United States Army station.

3. The Provost Marshal, A.F.G., will take the necessary steps to make known these instructions to the members of his organization and to the German police authorities.

4. All commanding officers will promptly, upon receipt of notice from a German or American official that any person belonging to the American Forces in Germany, cause a Medical Officer to examine, without delay, the injured party and submit the report directed in Paragraph 1 of this order.

APPENDICES

1. Policy Relative Illegitimate Children.
2. "Strained" Relations.

- - - - - - - -

APPENDIX No 1.

POLICY RELATIVE TO ILLEGITIMATE CHILDREN.

The high rate of exchange, etc.
(see attached article)

APPENDIX No. 2.

"STRAINED" RELATIONS.

What was said at the beginning, etc.
(see second attached article)

CHAPTER I

1. The high rate of exchange for the dollar, and the scarcity of German men of marriageable age in the American occupied area, early led to the establishment of illicit relations between soldiers of the American occupying forces and certain types of German women. This relationship did not assume an aspect of importance until after the lifting of the ban upon fraternizing during September 1919. Early in 1920, Mr. B. O. Mayer, Regierungs assessor, sent the following letter to the Officer in Charge of Civil Affairs:

"Subject: Illegitimate children of members of the American Occupation Army.

Sir:

I beg to draw your attention to the fact, that a great number of illegitimate children is living in the Landkreis Coblenz, whose fathers belong to or have belonged to the American occupation army. It will scarcely be possible to ascentain the number exactly, as many mothers, as may easily be understood, are in dread to state, that a member of the American occupation army is the father of her child. Nevertheless it will not be fixed too high, if from former statements the number of these children is put now at 200-300.

According to the German laws (Section 1708 and following of the Civil Code) the father of an illegitimate child is bound to grant him until to the termination of its 16th year a livelihood according to the position of the mother, which includes the expenses for the whole wants of life as well as of the education and of the professional training. The obligation to pay alimony for the illegitimate child passes the completed 16th year, if the child at the attainment of this age is in consequence of a mental or physical defect not able to make a livelihood. According to Section 1710 of the civil code this livelihood must be granted by an income in money which, however, according to #1714 of the civil code, if the father and the guardian of the child agree about it, can also be converted into capital for indemnity with the consent of the court for the protection of the wards.

However members of the army of occupation who are fathers of German illegitimate children cannot be sued by the guardians before the German courts of justice for payment of a livelihood. Whether or not they are willing to provide for the needs of the children, which they have begotten, is entirely left to their discretion as long as the present legal procedure is going to last. According to the reports of the nurses of the district, who have also to attend to the care of the illegitimate children, it must be appreciated that many members of the army of occupation are going to marry the mothers of their children and by this fulfill their moral duties toward the mothers and children. Others also assist the mothers partly after their discharge from the service in the American army and their return to America with considerable contributions in money. Some cases have also become known in which the parents of a soldier from America have made assignments of money, food-stuffs and articles of clothing to the mother of the child, who has been begotten by their son in Germany.

The assistance which is given to the children in all these cases depends only on the free will of the soldiers concerned. It can at any time be stopped again and is going to be discontinued in many cases upon the return of the soldiers to America. In a great number of cases the soldiers withdraw however entirely from their moral liabilities to the child and do not make any payment for the maintenance as long as they belong to the Army of Occupation relying on the fact that there is no possibility to constrain them according to the present laws.

It is quite evident, that by this a great number of these children, especially at the great dearness which is existing in the occupied territory, become destitute and are abandoned to a pitiful condition. The mother of the illegitimate child, who has then in the first instance and ofter alone the duty for the maintenance of her child, is very often not able to look for a lucrative employment. If also, as happens very often, the parents of the mother are dead or are living in needy circumstances and cannot pay for the maintenance of the child, so in the end it remains only to make application to the public poor-law administration.

Under which dreadful circumstances these poor children are growing up, who themselves are without guilt in the slip of their parents and of the political conditions, needs no further explanation. It would only answer to the general human fellow-feeling, which the American authorities have manifested till now in such magnanimous way to the German children, if by their assistance a way could be found out to secure the maintenance of the children whose fathers are members of the occupation army till to the return of normal conditions. I beg to submit the following propositions with the request, kindly to examine them and if it is in your power, to support them. To the Welfare's office for the Landkreis Coblenz a professional guardianship has been attached since some time. An official of the welfare's office will be appointed by the German courts of justice as guardian of all illegitimate children in the Landkreis Coblenz. To his duties belongs in the first line, to investigate what the father of the child is. If it proves that an American soldier is coming in question as father of the child, his name and organization should be ascertained. I will then submit all papers to you and ask you to summon the soldier and if possible to induce him to own the child directly before you and to pay spontaneously a certain amount to you for the maintenance of the child. You would then assign the amount to me and I would invest the money for the child in the Kreis-Sparkasse in such a manner, that money could only be drawn with my written approval. The payments which are made in some cases by the American soldiers to the mothers of the child fail very often in their purpose according to the reports of the nurses of the district, if the mothers are light minded persons, who don't spend the money for the child, but for themselves. This I will prevent by the investment of the money in the above described manner.

By the acceptance and performance of my proposition I think that at least in many cases a guarantee for the maintenance of the children can be obtained, who are without guilt in their position.

I ask further for your assistance in those cases in which it is the question to ascertain the residence of American soldiers who have already been discharged to their native place, to induce them if possible to the performance of their duty for the maintenance. In the interest of the poor children, I should be very much obliged to you for an early answer, if you think, that my propositions are practicable and if you will give in principle the consent of your assistance.

Respectfully,

B. O. Mayer, Regierungsassessor."

This letter received careful consideration and the officer in charge of Civil Affairs, in his 4th Indorsement of June 22, 1920, set forth the policy governing the question of illegitimate children--

1. Returned.

2. The question of establishing a policy with respect to illegitimate children of members of the American Army of Occupation has recently arisen in connection with an official communication received from the Landrat, Landkreis, Coblenz. There is a copy of this communication hereto attached. This is the first official communication which has been received from the German authorities on the subject.

3. The matters contained therin, having been admitted for his consideration and decision, the Commanding General directed that the policy shoudl be as follows, namely:

That all American military authorities can do at the present time is to assist in locating the soldiers who are responsible in these cases, and having done so, to point out to those culpable, through their respective Commanding Officers, the moral obligation incumbent upon them to provide for mother and child. The payment of funds for this purpose would have to be a voluntary proposition on the part of the soldier involved, and in case he should admit his responsibility, would have to pass direct from the soldier to the woman concerned in accordance with a private agreement, i, e., the office of Civil Affairs would not act as a collecting agency nor as a depository for funds of this kind.

4. The difficulty of establishing a general policy to fit all cases is apparent. In many cases the soldiers have returned to the United States and have been discharged from the Military Service. In other cases soldiers are still in the service in the United States, but deny responsibility and will endeavor to show that the woman was a prostitute. Here in Germany, the situation is complicated by the fact that at the present time we do not permit soldiers to be sued in, or prosecuted before, the German courts.

5. Hence, it is the opinion of this office that the policy above outlined is the best general policy that can be followed and that each case, after thorough investigation, will have to be decided by the Commanding General on its merits before bringing moral pressure to bear on the particular soldier accused.

6. Under date of 21 June, 1920, the Landrat, Landkreis Coblenz, was informed to the foregoing effect.

H. S. GRIER, LT. COLONEL G.S.
Officer in Charge of Civil
Affairs."

The German law upon this subject, which is set forth below in a paper dated September 20, 1920 and signed by Herr Verstegen,

Secretary of the Amtsgericht, was used as a guide in following this policy

SUBJECT: Relation of unmarried fathers to the illegitimate children and the mothers.

According to Section 1708 of the Civil Law the father of an illegitimate child is obliged to support the child until the completion of its 16th year. Section 1708 reads as follows:

"The father of an illegitimate child is obliged to support the child until the completion of its 16th year according to the station in life of the mother. This support includes necessaries of life, common education and preparation for a profession or trade. If upon arriving at the completion of the 16th year the child can not support itself due to physical or mental incapasity the father is obliged to support it beyond the time mentioned." Therefore, the period that the father is obliged to support the child may extend in special cases beyond the 16th year. However he is not obliged to support the child if due to other obligations he is not in a position to so do without endangering his possibilities of meeting other obligations. As, for instance, should he have a family of his own to support this family must first of all be provided for, only after so doing is he to support the illegitimate child.

The manner of support as specified by section 1710 of Civil Law is as follows:

The support is to be provided for by money paid in advance. This payment is made in periods of three months in advance. These payments do not release the father from future obligations. If the child is alive at the beginning of the next quarter a full instalment for the following quarter must be paid.

This support may also in accordance with section 1711 be demanded for past periods. The right of claim does not expire at the death of the father. The obligation still exists even if the father died before the birth of the child. In this case support can be claimed against the estate of the father.

The right of claim for the support of the child expires at the death of the child provided there are no past payments due.

In the case of death of the child the burial expenses must be paid by this father if payment therefor can not be obtained from the heirs of the child.

The father and child, the latter represented by a guardian may settle upon a lump sum. This agreement, however, is subject to the approval of the guardian's court. A final reaunciation for future support is not admissable.

As far as the support itself is concerned there are no fixed individual circumstances. The standard is the Whereas formerly a minimum of Marks 45 per quarter appeared sufficient under pres ent conditions marks 225 per quarter or marks 75 per month must be considered a minimum. It is, therefore, customary that the legal courts who state the amount at from marks 225 to marks 270 quarterly in ordinary cases, domestics, etc. Juvenile asylums who generally take care of these children demand on an average, marks 90 to 95 monthly. An increase of these figures for support quarterly appears logical owing to the present high cost of living conditions.

The position as to the obligations of the unmarried father are arrived at in the following manner:

(a) If the father admits that during the legal receptive period which is from the 181st to the 302nd any before the birth of the child that he had sexual intercourse with her and that he does not claim that during this period she had intercourse with other men he must make such a statement in court or before a notary admitting that he is the father of the child and acknowledges that he is obliged to pay a certain sum, payable quarterly in advance, up to the 16th year. In this statement he makes himself liable and may have brought against him a civil suit as debtor in the courts.

(b) If he denies having had sexual intercourse with the child's mother during this receptive period or if he claims that she had intercourse with others during this period suit may be brought in court to prove the statements." If the legal representative of the father who acknowledges the child or who is as found to be by the decision of the court agrees to a settlement it must be made a matter of court in record or done by affidavit taken before a notary. This affidavit, however, and agreement also is subject to the approval of the guardians' court. In all cases the rights of the child must be protected.

In arriving at the amount of money to be paid the following conditions prevail. If the father is able to pay the amount of settlement the interest added for the period of 16 years under conditions existing today must be at least Marks 12,000 to Marks 15,000. This amount is deposited with the guardians's fund by the guardian, the book of deposit being so made out that payments can not be made for any other purpose than so arranged. The support of the child is then paid from the interest in so far as this is found sufficient. Otherwise the amounts in excess are charged against the principal.

The father of the child is hereby released from further payments.

If the whereabouts of the father of the child are unknown and if the mother has during the receptive period mentioned in the Law had intercourse with other men the child is dependent solely for its support upon the mother. If she is not able to earn sufficient to support the child there is nothing else to be done but that she or the guardian of the child apply to the department of the poor in the locality of which she is a resident, which department will decide upon an amount to be granted her, or that the child be placed in some asylum at the expense of the Gemeinde. The officials of the Department of the poor, however, will only act on request and after careful investigation. If the mother is married to some other than the father of the child the father is not released from his obligation to support.

If the father of the child marries the mother, the child acquires the full rights of a legitimate child.

SUBJECT: The obligations of the child's father to the child's mother, is provided for as follows in Section #1715 of Civil Law.

The father is obliged to pay the costs of confinement as well as support for the first six weeks after the birth of the child. If during pregnancy or confinement other charges are necessary they must also be met by the father. The amount the mother may charge for these costs may be without regard to the actual costs. The right to this claim still exists even if the father of the child died or the child was still-born. The status of limitations in this case is four years. This authority begins six weeks after the birth of the child.

The claim of the mother of the child, however, does not affect the rights of the child. The Guardians' Court is guided entirely in the matter of support by the wishes of the mother or her representative. Any suit begun or any settlement under descussion is not subject to the approval of the guardians' Court.

Even before the birth of the child a request by the mother may be submitted that the father of the child shall make payment for the first three months beginning with the birth of the child, to be paid to the mother or the guardian within a reasonable time before the birth of the child. These means which are little known are very rarely resorted to.

Coblenz, September 30, 1920.

(Signed) VERSTEEGEN,
Secretary of the Amtsgerichts

2. Various cases of illegitimate children were handled

and the money collected from the man in the case was turned over to the German official guardian for illegitimate children to be disbursed for the benefit of the child and its mother. In some cases where the facts in the matter were in dispute, the German woman alleged to have been wronged, was permitted to sue the delinquent soldier in a German court in order that the case might be decided upon its merits.

- - - - -

APPENDIX NO. 2

"STRAINED" RELATIONS.

What was said at the beginning of this chapter with regard to the relations between the A.F.G. and the civil population was subject to the exception of one period of time when the attitude of the Germans became comparatively unfriendly.

In the summer and autumn of 1921 a change in the general attitude of the Germans became apparent, and at the same time there occurred a marked increase in the cases of improper conduct by American soldiers. Complaints by Germans became more frequent and the press in unoccupied Germany published articles in sharp criticism of the American occupation.

Cognizant of the increasing number of offenses against Germans that were being committed by American soldiers, the Chief of Staff on October 11, 1921, sent the following letter to brigade, regimental and separate unit commanders:

> 1. Offenses committed against Germans by soldiers of this command appear to be on the increase in spite of all the efforts of Commanding Officers and others concerned to bring the

offenders to justice for such offenses. These offenses usually take the form of assaults by soldiers against German and so many of such cases have occurred as to seriously jeopardize the good name of the American Forces in Germany.

2. Many cases of assaults against Germans where it appeared from a reading of the reports thereon that there had been a serious miscarriage of justice, and that too great leniency had been shown the offenders, have been ordered re-investigated by this office. The reports in almost every case on re-investigation have disclosed a tendency on the part of officers either to shield their men from the results of their misconduct or to view with too much leniency the offenses for which their men are being tried. Many officers appear to regard an assault against a German as a comparatively trivial offense and in many cases have punished the offender under the 104th Article of War. This attitude of the officers of this command is believed to be primarily responsible for the great number of cases of assaults against Germans.

3. In reference to this condition of affairs there recently appeared in the "Germania", a newspaper published in Berlin, a letter written by the Coblenz correspondent of the paper, an extract from which reads as follows:

> "The American authorities could, if they desired, proceed very drastically against this undisciplined and barbaric behavior of their soldiers, primarily by investigating thoroughly all cases reported and by meeting out summary punishment to the guilty. It probably holds true that an indirect cause of these assaults is the mild series of sentence issued by the American courts."

4. Commenting upon the above extract from the "Germania", the Commanding General, these forces, has the following to say:

> "It has come to my knowledge that a few officers of this command seem to regard with a certain indifference an incorrent attitude and deportment at times of themselves and their commands towards the inhabitants of the Rhineland. This viewpoint is wholly unintelligible to a justly thinking mind and far worse under the conditions of an occupying force than it would be under normal ones. It is desired that those who may harbor such thoughts or indulge in such an attitude make an introspection with a view to making a change that will enable them to carry out what they should well know is the policy of these Headquarters; namely, consideration and courtesy at all times.
>
> HENRY T. ALLEN,
> Commanding General."

5. All officers of this command are reminded that subordinates, whether they be officers or enlisted men, take their cue from their superior officers and that when those concerned with the enforcement of discipline show a tendency to regard as trivial offenses committed against the person and property of Germans, and to protect and shield their men in the commission of such offenses, such an attitude is at once communicated

> to their subordinates, and must inevitably result in the breakdown of discipline. It should be the concern of every officer and enlisted man of this command to guard and protect the reputation of the American Army in Germany, and in no way can that better be accomplished than by letting all concerned know that we stand for courtesy and fair-dealing for all, whether they belong to the nations of our late enemies or not.

This was followed on October 12 by a memorandum to all officers, quoting the above remarks of the Commanding General for their information and guidance.

No change in the attitude of the Germans was noted, and in November an extended investigation of the subject was made by G-2. It was found that the attitude of the Germans, while far from hostile, had undergone a considerable change as compared with a year previously. There were many factors which had served to bring about this change. Some of them had to do with the successive disappointments of the Germans who had expected the United States to "save Germany from the revenge of the Allies." It seemed probable that a minor propaganda was going on in the American area, not political but rather "put forth in the hope that thereby some alleged abuses in the American zone will be brought to the attention of the proper authorities and so will result in an amelioration of the conditions complained of."

It was the belief of G-2 that there were two main causes of the existing German attitude. First and most important were the numerous cases of misconduct by intoxicated American soldiers. The congested billeting situation and "the harsh attitude of some Americans towards the German householders" were the next most prolific sources of complaint. The Germans asserted that the American courts were too lenient towards Americans charged with offenses against Germans and too harsh when a German was the defendant. The official class primarily, and secondarily the householders and middle classes in general, complained "that the Americans, particularly the officers, are either domineering or condescending toward them, and that their attitude is more rigorous than the circumstances warrant. This feeling would probably exist in any case, but it is probably true that some Americans unnecessarily humiliate self respecting Germans." "There are many minor factors such as the low value of the mark and the high pay of American officers and soldiers, the 'luxurious' life of Americans in German residences and clubs, the 'cleaning out' of Coblenz shops by American women and the coincidental rise in prices, the presence in Coblenz of so many discharged soldiers and other American civilians (some in occupations of a more or less queer nature), etc., etc." The conclusions of G-2 ended as follows:

> In general it is believed that the reasons for the change in the feeling of the inhabitants are to be found in the actions of the American troops of occupation. Many of the matters complained of are inseparable from an occupation, but many are entirely separable therefrom. It is the latter that must be corrected, not because of what the Germans may think of us but because of our own self respect and of the good name of our country.

It was apparent that the situation called for remedial action, as the misconduct of the American soldiers constituted a reflection upon the high reputation of the A.F.G. In early December the Chief of Staff and the Officer in Charge of Civil Affairs held conferences which all officers were required to attend and at which all phases of the situation were discussed and directions as to remedial action were given. On December 15 the Chief of Staff addressed the following memorandum to brigade, regimental and separate unit commanders:

> 1. Enclosed herewith is a report on the above subject made by the A. C. of S., Intelligence. The report is based on an extended investigation and copies of supporting papers are attached.
>
> The report, with attached papers, is furnished for the careful study and guidance of all concerned.
>
> 2. On this subject, the Commanding General remarks as follows:
>
> > "Our policy here must for every reason of justice and righteousness be founded on scrupulously correct conduct towards all inhabitants of the Occupied Territory."
>
> 3. The policy of the Commanding General towards the people of the Occupied Territory is well known to the personnel of these forces. The keynote of that policy is courteous treatment of the inhabitants of the Occupied Area in all our official and un-official relations and even handed justice in the administration of the provost and other military courts, - in other words, a square deal for all.
>
> Now that America is at peace with Germany, there is all the more reason why this policy should be continued and broadened.
>
> 4. It is the desire of the Commanding General, that you cause all officers of your command, to read carefully this letter and all accompanying papers and that you acquaint all enlisted men of your command with the gist of so much of G-2's report as refers to the misconduct of American soldiers.
>
> You will be held responsible that the attitude of the officers and men of your command conforms in future to the Commanding General's policy of fair treatment of, and courtesy and justice to all inhabitants of the Occupied Area.

This action and the steps taken by the commanders to whom the memorandum was addressed had an immediate effect. There was a marked decrease in the cases of misconduct by American soldiers and in the complaints by Germans, while a change in the attitude of the Germans quickly became apparent.

The "strained relations" were strained only as compared with those that had existed before the autumn of 1921 and were not actually so in the general acceptation of the term. At the very time (in the autumn) the minor propaganda mentioned was at its height and complaints were most numerous, the Germans repeatedly expressed their grateful appreciation of America's failure to participate in the various measures against Germany, of American efforts (attributed almost wholly to General Allen) with the Interallied Rhineland High Commission, and of the fact that the United States had no selfish ends in view. The propaganda was clearly not political, and the Germans often expressed their keen desire to have American troops, though at reduced strength, in the Rhineland as long as any troops of occupation remained there. And the Germans frequently voiced their realization of the fact that it was but a very small minority of the Americans whose conduct or attitude furnished grounds for legitimate complaint.

During the period mentioned life in the American area went on in practically the same manner as before. The soldiers and their German friends, mostly of the gentler sex it is true, continued to associate to the usual extent, the misconduct of the American soldiers and the cases of individual friction almost invariably occurred under the cover of darkness, and visiting observers continued to comment upon the numerous evidences of friendly relations. So, while the situation was one that undoubtedly required corrective measures, it should be viewed in the light of the foregoing, with the realization that the "strained relations" were technically rather than actually so. In conclusion, it may be said that the Germans stated at this time that they felt "that if General Allen knew of their legitimate complaints he would quickly have them redressed".

CHAPTER NO. 5

SANITATION AND PUBLIC HEALTH.

Control of Sanitation; Prisons and Hospitals; Bendorf Hospital; Bathing Establishments; Barber Shops, Theatres, etc.; General Health Conditions; Contagious Diseases; Births and Deaths; Officers.

REFERENCE.

General . . . . . A.M.G., Volume I, Chapter 7.

SANITATION AND PUBLIC HEALTH.[1]

The activities of the Department of Sanitation and Public Health were not greatly affected by the transfer of Civil Affairs duties to the American Department of the High Commission. All orders and regulations concerning sanitation and public health were continued in force, and in most cases were still in effect on December 31, 1921.

The principal duty of that department was to control the sanitation of the American area with an ultimate view to the protection of the troops from disease. This control was exercised through the Medizinalrat (medical adviser to the Regierungspraesident) and his subordinates, the various Kreisarzte, all of whom co-operated to the fullest extent.

The control of sanitation involved the following:

> Weekly inspections of German prisons and hospitals (including the American military prison, where Germans were confined).
>
> Inspections of public bathing establishments.
>
> Inspections of barber shops, cafes, hotels, theatres, cinemas and slaughter houses.
>
> Weekly reports of contagious diseases in the various Kreise.
>
> Monthly reports of deaths in the Kreise.
>
> Monthly reports of births - legitimate and illegitimate, including cases of alleged American parentage.
>
> Liaison and co-operation with the German Fuersorge (Welfare Society) in connection with juvenile offenders.

PRISONS AND HOSPITALS.

Germans convicted of violations of military orders or ordinances of the High Commission and sentenced to imprisonment were usually confined in the Kreis jails. These prisons and the German hospitals

[1] Volume I.

were always maintained in a very good state of sanitation, and overcrowding was prevented. During 1920 and 1921 there was no epidemic of respiratory or intestinal diseases in any of the prisons or hospitals - an interesting fact in view of the widespread epidemic of influenza that occurred in December 1921.

In the jails the food was the ordinary prison fare and contained all the necessary elements of a balanced diet. There was no evidence of nutritional disturbances, while many inmates gained in weight. Each prison had its own infirmary for the treatment of minor illnesses; serious cases were transferred to the Burger Hospital in Coblenz.

In 1920, arrangements for the confinement at Andernach of prisoners sentenced by Superior Court No. 2 (2d Brigade area) having been made, the hospital at Niedermendig was released from American control and turned over to the German authorities.

Bendorf Hospital

During the World War many German troops had been temporarily stationed in the Rhineland, and there had been constant troop movements through that section of Germany. As a result the venereal rate among the women inhabitants had become exceedingly high. The American forces of occupation found themselves confronted with a serious venereal problem, and measures to control the ever-in-creasing venereal rate were put into effect as soon as practicable.[1] There was an alarming increase in the A.F.G. venereal rate after the revocation of the anti-fraternization order in September 1919, and the Staatsarchiv building was chosen as a hospital for the

[1] A.M.G., Volume I, Pages 196 - 201.

treatment of women infected with venereal disease. Regulations establishing a court for vagrants (including prostitutes) and juvenile offenders were issued. After the adoption of these measures the venereal rate decreased slowly but steadily from the extremely high annual rate of 422.65 per 1000 (October 21, 1919).

By May 1920 the number of female venereal patients exceeded the capacity of the Staatsarchiv building. The Bendorf hospital was then selected, and on May 1 it began to function. At first the bed capacity was 250; this was increased later to 300. The hospital was operated with the following personnel (46 in all):

Chief Doctor	1
Doctors	4
Nurses	16
Laboratory Worker	1
Managing Officials	2
Women Clerks	2
Guards	14
Cooks	2
Housemaids	2
Seamstress	1
Laundress	1

All expenses of the hospital were borne by the German Government, but it was operated under the supervision of the Department of Sanitation and Public Health (Civil Affairs).

994 patients, representing 28% of those convicted by the Vagrancy Court, were admitted to the hospital in 1920; during the following year 811, or 35% of those convicted, were treated there. It is to be remembered that only those women infected with venereal disease were admitted to this hospital.

Under the direction of Dr. Kurt Roscher of Berlin, the hospital maintained a high standard of scientific work. Wasserman, Meinicke and Sachs-Georgi reactions and other tests were made in the laboratory, and the most modern therapy was employed.

The operation of this hospital and of the Vagrancy Court achieved results that thoroughly justified the adoption of those measures. They were the largest factors in the protection of the soldiers from venereal disease.

BATHING ESTABLISHMENTS.

In each of the larger towns in the American area one or more municipal bath houses were operated for public use at a nominal charge. There were three such bath houses in Coblenz, one in Andernach and one in Mayen. All were operated efficiently and were exceptionally clean. These bath houses were used extensively by the American troops stationed in the respective towns.

In Coblenz the municipal swimming pool was requisitioned by the American authorities and was operated by the Y.M.C.A. for the benefit of the officers and soldiers. It was equipped with hot and cold shower baths, the use of which, before entering the pool, was compulsory.

BARBER SHOPS, THEATERS, ETC.

Barber shops desiring to cater to military trade were required to secure permission therefor from the Provost Marshal and to follow certain sanitary rules. Upon the receipt of an application for permission to serve Americans, the applicant's shop was inspected. If the results of the inspection were satisfactory, the barber was given a police permit which he was required to post in a conspicuous place. Thereafter the shop was inspected at intervals; if conditions were found to be below the required standard, the shop was placed "Off Limits" for a variable length of time. In addition to passing a satisfactory inspection, each barber was required to undergo a monthly physical examination by a German doctor, and when called upon to produce a certificate, signed by that doctor, that he was free from disease. In general the barber shops were operated in a satisfactory manner, but occasionally it was found advisable to place one "Off Limits".

The kitchens and ventilating systems, in particular, of cafes and hotels were inspected periodically. In general these were maintained at the required standard.

Inspections were made at intervals to see that theatres and cinemas were well ventilated and that there was no over-crowding. The results of these inspections were generally satisfactory. Ventilation was usually maintained by fans which produced a vacuum system. Occasionally a moving picture theatre was placed "Off Limits" for over-crowding. In December 1f 1921 a number of smaller theatres and cinemas were placed "Off Limits" on account of the epidemic of influenza.

The German system of meat supply is different from that used in the United States where large packing houses, located at a comparatively few points, ship refrigerated beef to all parts of the country. In Germany each town maintains its own abattoir, which supplies its needs and those of the immediate vicinity. The Coblenz abattoir was highly efficient. Every Tuesday beef and hogs were slaughtered, and immediately afterwards the meat was inspected by a veterinarian.

GENERAL HEALTH CONDITIONS.

At the beginning of 1920 the general health conditions among the civil population of the American area were very good, and with the exception of a few epidemics they remained so up to the end of 1921. The usual contagious and infectious diseases (measles, scarlet fever, diphtheria, tuberculosis, typhoid and para-typhoid) were always present, and occasionally they would break out in epidemic form. These were usually remote from the American troops and did not become a factor in their sick rate. The widespread influenza epidemic of December 1921, however, was undoubtedly the cause of a similar epidemic and a high respiratory rate among the troops.[1]

CONTAGIOUS DISEASES.

Under German law influenza and pneumonia were not reportable diseases, and no accurate figures concerning them could be secured.[2] The general health of the population was so good during 1920 and 1921 that, except as noted below, it was not deemed necessary to show the rates of incidence.

Sleeping Sickness. During January and February 1920 cases of this disease (Eucephalitis Lethargica) were reported from various points in the American area. Investigation resulted in a belief that none of these cases should have been diagnosed under that head. The German Government had issued a pamphlet on influenza, in which special attention to this form of sleeping sickness as one of its sequellae was directed; it was thought that the reports were due to this pamphlet.

Influenza. Among the Germans influenza is not a reportable disease. Ordinarily a German does not consult a doctor unless he

1 Volume II, Chapter 26.

2 Whenever influenza became an important factor in the general health situation, the fact was reported by the German authorities to the Department of Sanitation and Public Health.

is very ill; hence doctors probably would not hear of influenza cases until after pneumonia had developed.

In February 1920 an epidemic of influenza occurred in Kreis Unterwesterwald. Out of the 474 cases, 224 were in the town on Hoehr. In December 1921 there was a wide spread epidemic throughout the Rhinelands, including the American area.[1] While the influenza was of mild type, many cases of pneumonia developed. No figures, as to either influenza or pneumonia could be secured, but the former was very prevalent and at one time all beds in the German hospitals were occupied by pneumonia cases.

Para-typhoid. In April 1920 there were 50 cases of this disease in a boy's school at Kamperhof (Landkreis Coblenz). In August of the same year 67 cases occurred among the school children of Braubach (Kreis Neuwied), a village of 688 inhabitants.

Measles. 145 cases occurred in the American area during August 1920; 73 of these were in Kreis Neuwied.

Tuberculosis. Under German law this disease is not reportable unless death occurs; therefore the German reports do not furnish a true index of the incidence of this disease. In October 1920 all Kreisarste and other physicians were directed to make a tubercular survey of their respective Kreise and towns, and in future to report tuberculosis in the same manner as other reportable disease. The survey, made in October and November 1920, showed that 2.2 persons out of every 1000 were afflicted with tuberculosis. The highest rate was in the city of Coblenz, where there was an incidence of 6.4 per 1000.

[1] See Volume II, Chapter 26.

BIRTHS AND DEATHS.

Births. It was expected that after the signing of the Treaty of Versailles the German birth rate would quickly return to that of pre-war years. This came to pass in the American area. The rate for 1919 (12.60 per 1000) was the lowest since 1909. In 1920 the rate jumped to 26.38, the highest since 1911, and in 1921 the rate was 29.13 per 1000, which was the highest since 1909.

Deaths. With the exception of 1913, when the death rate was 15.07 per 1000, the rates in 1920 and 1921 were the lowest since 1909. In 1920 the death rate per 1000 was 15.74, and in 1921 was 15.70. In 1920 the birth rate exceeded the death rate by 10.63, and in 1921 by 13.41. These figures correspond with those for Germany as a whole, which are shown graphically (at the end of this chapter) from data furnished by the German Government.

Illegitimate Births. The following statistics, relating to the American area, are of interest:

Year	Illegitimate Births	% of All Births
1920	1134	9.5
1921	863	7.06

Of the illegitimate births in 1920, American parentage was alleged in 411 cases of these 159 or 38.7% were reported from Kreis Mayen. In 1921 American parentage was alleged in 365 cases, of which 34.5% occurred in Kreis Mayen. Assuming the statements of the mothers - which should be accepted under reserve - to have been true, the comparatively large numbers reported from Kreis Mayen might be explained by the fact that, with few exceptions, all soldiers in Kreis Mayen were billeted with private families, while elsewhere they were quartered in barracks.

Still Births. There were 322 still births in 1920 and 345 in 1921.

OFFICERS.

Lieutenant Colonel Walter Bensel, Medical Corps, organized and conducted the Department of Sanitation and Public Health until early in 1920, when he was relieved by Major (then Colonel) Morrison C. Stayer, M.C. In September 1920 the latter was replaced by Major Thomas J. Flynn, M.C., who served in that capacity throughout 1921.

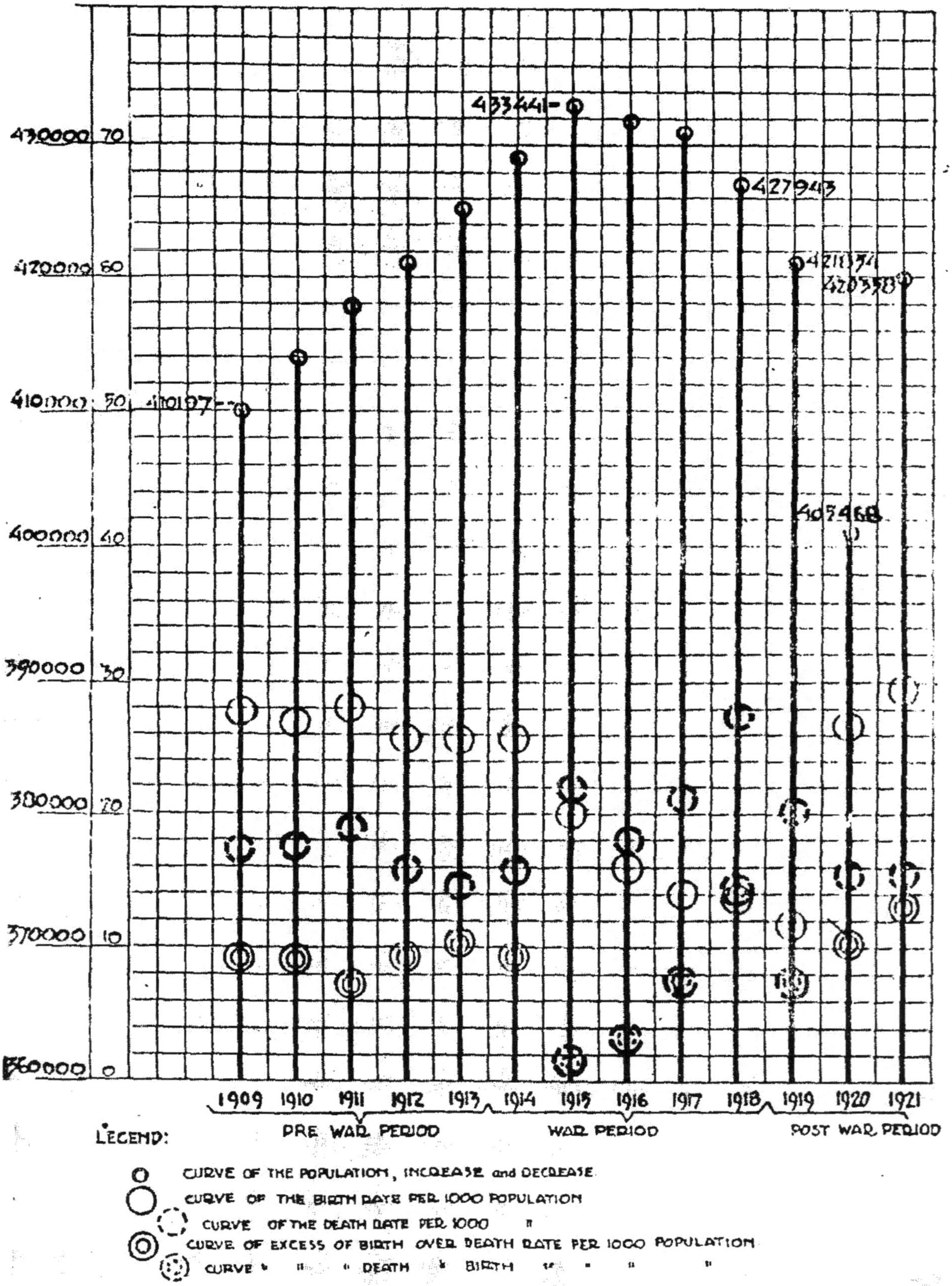

430000 70
420000 60
410000 50
400000 40
390000 30
380000 20
370000 10
360000 0
410197
433441
427943
421134
420558
405468
1909 1910 1911 1912 1913 1914 1915 1916 1917 1918 1919 1920 1921
PRE WAR PERIOD
WAR PERIOD
POST WAR PERIOD
LEGEND:
CURVE OF THE POPULATION, INCREASE and DECREASE
CURVE OF THE BIRTH RATE PER 1000 POPULATION
CURVE OF THE DEATH RATE PER 1000 "
CURVE OF EXCESS OF BIRTH OVER DEATH RATE PER 1000 POPULATION
CURVE " " " DEATH " BIRTH " " " "

CHAPTER NO. 6

PUBLIC UTILITIES.

General Conditions; Policies; Narrative; Roads; Officers.

REFERENCE.

General . . . . .A. M. G., Volume I, Chapter 9.

PUBLIC UTILITIES.

GENERAL CONDITIONS.

In January 1920 the public utilities throughout the American area were barely able to meet normal demands, and even ordinary requirements were not always satisfied. The needs of industry and private consumers were increasing steadily. The industrial activities of the area were stimulated by the rapid depreciation of the German mark, and there occurred a commercial boom which was more or less artificial and which was accompanied by a great rise in prices. The large profits obtained by manufacturers caused them to increase their output and in some cases to enlarge their plants. In consequence the demands upon public utilities, particularly for power, increased so rapidly as to exhaust whatever reserve producing capacity the plants may have had.

The presence of the troops of occupation involved an additional burden upon public utilities. The Americans were prodigal in the use of heat, light and water. Comparative statistics show the per capita consumption of these essentials to be much greater in the United States than in Germany. The A.F.G. proved no exception to the rule. Records demonstrated that the water consumed by units of the German Army in Coblenz in 1913 was but one-fifth of the average consumption of American units quartered in the same barracks. Wastefulness would partly account for this great difference, but it was mainly due to different standards as to bathing and toilet arrangements. The effect of the occupation upon the consumption of gas and electricity was not easily ascertained, but in every case where it was possible to analyze the demands, the requirements of the troops and of American and Allied families in billets were found to be in excess of those of the Germans. While the unit consumption of the A.F.G. was relatively high, it did not cause a large increase in the total demand. It was estimated that at most but a 5% increase in the total water consumption of Coblenz could be attributed to the occupation. In some smaller communities, particularly where before the war German troops had not been quartered, the effect was greater.

It would not have been difficult to supply the increased demands had it been practicable for public service organizations to have enlarged their facilities. But conditions following the war were so abnormal that while the demands of consumers were stimulated the development of public utilities was retarded. No extensions to existing plants had been made during the war, and all but the most necessary repairs had been postponed.

After the Armistice the rapid rise in the prices of machinery continued to discourage expansion. The nature of a public utility prevents it from using its current earnings to pay for an extension of its plant, and its profits are necessarily small in proportion to its turnover. Such a concern is very sensitive to public opinion in the matter of rates, and it hesitates to provoke popular discontent by rapidly increasing them. For example, after every increase in fare on the Coblenz street car lines, the number of passengers diminished.

Hence the power plants could not keep pace with the increasing demands of local industries, and they hesitated to multiply their overhead charges in order to meet a demand that at any time might return to normal. Frequently agencies that supplied gas and water were owned by a municipality which was too poor to appropriate large sums for extensions of public utilities. In 1919 some progress in rehabilitation had been made, but the condition of many of the more important plants was far from satisfactory. The plants supplying Coblenz with electricity and water were in particularly poor condition.

In spite of obvious limitations upon production, the managers of many public utilities, particularly the gas and electrical works, did not hesitate to contract to supply more than they could produce. Rival concerns ran new transmission lines to districts where previously electric current had not been available. In the smaller towns gas stoves and cooking ranges were lavishly installed, without objection from the plants which later were unable to carry the increased load.

Shortage of coal served to intensify matters, and at best its continued supply to gas and electric works was uncertain. As the fuel required by the Allies for reparations consisted principally of the better grades of bituminous and gas coal, gas works were forced to use inferior grades. This increased their difficulties, for in order to generate the same amount of gas a proportionally larger amount of inferior coal had to be used. Electric power plants in turn were compelled to use brown coal (lignite).

The government agency for the distribution of coal (Reichskohlenkommissar) determined the allotment of fuel to various consumers and the order of priority in shipments. The limitations imposed were severe, and many plants were unable to operate effectively with the amounts apportioned. Constant pressure was necessary to obtain coal for garrisoned towns. The requisition from river barges of coal consigned to points in unoccupied territory was frequently resorted to, and the Quartermaster was often called upon to furnish small amounts of coal from Army stocks to various plants which otherwise would have been forced to shut down. Coal was shipped from the United States, but as the cost in the constantly depreciating Germany currency was too high for local consumers, the practice was discontinued and the American coal was issued to the troops.

POLICIES.

Due to the unsatisfactory conditions a constant and active supervision of public utilities continued to be necessary. As qualified technical personnel was not available elsewhere, this supervision remained a function of the Chief Engineer, one of whose assistants was designated as Officer in Charge of Public Utilities. The Officer in Charge of Civil Affairs formulated the general policies and was consulted when important decisions were required.

The responsibility for the active operation of the plants was put upon the German managers, who were required to execute all orders issued by the Officer in Charge of Public Utilities. The needs of the A.F.G. were always paramount; the demands of communities and industries were complied with as far as practicable, but only after those of the troops had been met.

Where existing funds were insufficient to cover alterations or improvements made necessary by the presence of the troops, a written order to perform the work was given to the responsible authority. Using the order as evidence, the community or private undertaking could then claim compensation from the German Government.

The issue of such orders was frequently necessary in small communities; the larger towns were usually farsighted enough to meet the wishes of the A.F.G. more than half way.

In adopting measures to control consumption, as was often made necessary by limited production, it was directed that the actual needs of the troops and the Allied community were to be satisfied in full and that all restrictions should fall upon the German population. The effect of this policy was to cause the plant managers to take more energetic steps to remedy conditions than would have been the case if the shortage had been shared by the Army. At the same time everything possible was done to avoid waste by Americans and Allies and to prevent their demands from becoming excessive.

NARRATIVE.

The year 1920 opened inauspiciously with a second flood of the Rhine, which surpassed that of December 1919. Traffic was interrupted and serious difficulties were caused to all public utilities in Coblenz. The situation was particularly bad at the water works on Oberwerth Island. There the flood water seeped into the pump pits, and one by one the pumping stations were put out of service. Finally but two large centrifugal units were able to continue in operation. Engineer personnel used army pumping machinery in the endeavor to prevent the water from making headway inside the pumphouses, and a fire engine belonging to the city of Coblenz was pressed into the same service. On January 15 the flood commenced to subside. Had the water risen six inches higher, the Diesel engines driving the last two pumps in service would have been forced to shut down.

The effect of the flood upon the gas works was limited; however one of the mines became filled with water which prevented the flow of gas. In the electrical works some transformers and transmission lines were put out of commission. Headquarters Building and the Coblenzerhof were connected to unaffected circuits on higher ground.

The winter and spring of 1920, similarly to 1919, were characterized by a shortage of coal, by strikes and by breakdowns in public utility plants. During March and April some of the pumping machinery at the Coblenz water works failed to function. The windings of the electric motor in Pumphouse No. 4 burned out because of the wetting that had been received during the January floods, and the large Diesels in House No. 3 broke down at about the same time. The amount of water available for Coblenz was materially reduced, and efforts to limit consumption by A.F.G. orders did not produce appreciable results. It followed that in some parts of the city the water pressure was low and the service irregular.

Serious inconvenience was caused when in February and March strikes occurred at the Hohm power house, from which the Coblenz Street Railway Company secured about half the current distributed over its lines. When this quantity was no longer available, a large number of the distribution circuits had to be cut off in order to prevent the load on the local generating station from becoming excessive. The reduction effected by eliminating all outlying and ungarrisoned towns proved to be insufficient; hence, in order to secure a further lessening of demand and to conserve coal, the suspension of the street car service was ordered. This measure proved effective; a settlement was quickly reached and work was resumed within two days.

The dry summer of 1920 caused a water shortage in the communities on the right bank of the Rhine in the vicinity of Ehrenbreitstein. These villages, as generally throughout the American area, were dependent upon springs for their water supply. The greatest difficulty was at Arzheim, where the diminished amount of water yielded by the local collecting works was not sufficient for the combined requirements of Fort Asterstein, the towns of Ehrenbreitstein and Niederburg and the community itself, all of which were supplied from the same source. When the water-supply system was built the funds were furnished by the German military authorities and the communities of Ehrenbreitstein and Niederburg. In return for this consideration, a certain quantity of water was to be furnished by Arzheim to the fort and to the other two communities. The situation was relieved by depriving Ehrenbreitstein (which had other sources of water supply) of water from Arzheim, diverting to Fort Asterstein the water thus made available. Ehrenbreitstein received compensation from another spring, originally used for the supply of Fort Ehrenbreitstein but no longer suitable for that purpose on account of the low gravity head.

In this connection it is interesting to note the development of water supply standards as illustrated by the installations in that fort. The ancient spring, which is elevated only a few feet above the level of the fort, supplied water to a series of cisterns located throughout the interior, and from these cisterns water was pumped up by hand as needed. The modern water-supply system had a pressure sufficient to cause a flow to all parts of the fort.

During the autumn of 1920, efforts were made at the Coblenz electrical works to increase the size of the plant in order to take care of the winter peak load. A 2000 KW steam turbine and correspondingly large boiler units were ordered. Work of installation was started, but it was not completed until the following spring. Keeping down overhead costs continued to be the most important factor, as was shown by the fact that old boilers and a second hand turbine of obsolete design and low efficiency were secured.

In November 1920 an interesting development was the co-operation between the Plaidt and Augustenthal electric generating stations. The coal reserve of the former having been reduced to a dangerously low point, the two works were synchronized and operated in parallel for several days. By this means the plant with the greater coal supply was able to take over a considerable portion of the other's load, and no limitations upon service became necessary.

A plan of restrictions was drawn up for use in Coblenz should conservation of coal or other considerations make it necessary to diminish the load on the local power station. The city and outlying districts were divided into zones, and a schedule was worked out so that restrictions could be rotated from day to day. Thereby less hardship would be caused to those consumers who had borne the brunt of the curtailments at some previous time when it had been necessary to diminish the load. The principle that the barracks and billets should be the last affected remained unchanged. It was not found necessary to bring the restrictions into effect until after the Christmas holidays, when the reserves of coal at the Hohn plant had reached such a low point that the production of that plant had to be cut down for a brief period.

Coblenz had always been discontented with the distribution of current from the Hohn central station. Before the war the city had concluded with the Coblenz Street Railway Company a contract by which a monopoly in the distribution of electric current had been awarded to that firm for certain considerations, among them the construction at Hohn of a modern generating station which was to be more than sufficient to supply the demands of Coblenz. The plant was built as required, but the demands of consumers soon outstripped its capacity. Besides the lines of the Coblenz Street Railway Company, this work supplied power to companies in the vicinity of Mainz and Frankfurt and furnished current to the large mining and industrial district of the Siegerland (Northeast of the Coblenz bridgehead).

When the shortage of power in Coblenz became emphasized by the reduction in the amount generated at Hohn, efforts were made through the Rhineland High Commission to secure a larger allotment of the output for the American area. In January 1921 a meeting at Hohn of the plant directors and representatives of consumers was arranged by the local delegate of the German Coal Commission (which regulates the distribution of electric current as well as of coal). A representative of the High Commission and the Officer in Charge of Public Utilities attended in unofficial capacity. The net result of the meeting was that conditions were found to be the same in the case of all consumers, and no one was willing to relinquish any portion of the power allotted to him.

Minor difficulties on transmission lines were continuous

throughout the winter. A typical instance, and one of the most serious, was the low veltage available at Fort Ehrenbreitstein. The transformer supplying the fort supplied the town also, and it was badly overloaded. During the war steel transmission lines had been substituted for copper, and the wiring in the interior of the fort was in very poor repair. After investigation a new transformer was installed and a copper transmission line restored, while a thorough overhauling was given to the distribution wiring. As the voltage drop was still excessive, lamps of a 10% lower voltage rating were installed.

The coal supply during the winter of 1920-1921 was better than before, but it was still much of a hand-to-mouth affair. In early December the Quartermaster was made responsible for the procurement of coal for public utilities.. The Engineer Officer continued to compile information and to furnish periodical statements of the amount of coal required. In order to secure better co-ordination with the German Coal Commission and more definitely to place the responsibility upon that organization, the Engineer Officer formulated a new policy which required the Coal Commission at the beginning of each month to submit its proposed allotments to works serving garrisoned towns to the American Headquarters for approval. These allotments could then be checked against the past performances of the plants concerned, and any changes necessary in the interests of the forces of occupation could be ordered by the Quartermaster upon the recommendation of the Engineer. Further, the amounts in tons of the "thirty days reserve," required to be maintained at each plant, were definitely established.

The success of this scheme was marked from the outset. Better co-ordination with the German authorities secured deliveries that were more satisfactory than those in the other occupied areas. The system stood the test of a complete suspension of rail shipments from the Ruhr District, which lasted more than a week during the railway strike in February 1922. This strike occurred very soon after a long period of a practical standstill of river traffic due to low water. Deliveries of coal from the Ruhr did not commence to come by rail until about 15 days after the beginning of the strike. In spite of the protracted interruption in deliveries, but three works had stocks under the required minimum reserve, and the average amount was well over 25 days' supply.

During the summer months of 1921 extensive alterations were carried out at the coking plant of the Concordia Huette steel works in Bendorf. The amount of gas available for delivery to Coblenz was reduced to less than half of the normal supply, but it was found possible to overcome the shortage by increasing the output of the local works.

The long dry spell of 1920-21 was sharply felt during the summer of the latter year in many parts of the American area. The Rhine fell to its lowest recorded leve, and the ground water table throughout the region sank correspondingly. Refinements at the Coblenz water works, in the nature of a vacumum system for priming the centrifugal pumps and preventing them from becoming air-bound, enabled this plant to give service in spite of the high suction head. Many of the springs, upon which nearly every town in the area relied, commenced to dry up early in the season. Most of the collecting works were located at shallow depths, without due consideration of the geological features of the neighborhood.

In fact the practice seemed frequently to have been to select sites for wells with the aid of diviningrod experts, many of whom still practice their craft in Germany. Methods for improving existing conditions were investigated at Mayen, Mulheim, Andernach and Ehrenbreitstein. The work was systematically carried out with the help of a geologist familiar with the area, and in every case the results obtained were satisfactory. Mayen presented the greatest difficulty and contained the most interesting features. Water for this town was obtained from two hundred meters of collecting galleries driven into a porous volcanic formation, saturated with water, which was located about 17 kilometers (10.6 miles) away. After a thorough preliminary examination the existing collecting works were deepened, and in order to obtain sufficient fall a portion (several thousand feet) of the aqueduct was reconstructed. As a relief measure until the project could be completed, a large centrifugal pump was temporarily installed in the main gallery.

The end of 1921 found public utilities in much better shape than before. The insistent demands for more power had led the Hohn central station to install a fourth turbo-generator unit of 4500 kilowatt capacity, and the Coblenz Street Railway Company had completed the setting up of the boilers required for operating their 2000 kilowatt turbine. Increased steam producing capacity was also available at the Plaidt and Augustenthal works. A project was on foot to connect the Coblenz and Plaidt works and so to permit exchange of power in an effect to secure better service. Throughout the area the water supply was adequate. Due to increased deliveries from the Concordia Huette, the gas supply of Coblenz was excellent. Provided coal deliveries continued to be satisfactory, no serious difficulties in the operation of public utilities was apprehended.

ROADS.

After the disbanding of the Third Army, the construction and maintenance of roads was turned over to the German Road Service. The Chief Engineer continued to supervise the work in so far as it affected military interests. The general policy was that roads required for army use should be maintained in excellent condition. With this end in view, road programs were co-ordinated and such mechanical aid as could be spared by the A.F.G. was lent to the German authorities.

In January 1920 there were twelve Liberty cargo trucks, one tank truck and one steam roller operated by German personnel under this arrangement. Fuel, lubricants and all repairs were furnished by the army, while the expenses were charged to the German Government as a part of the costs of occupation. The system proved highly satisfactory; the trucks rendered good service and the condition of the main roads in the American zone was generally acknowledged to be the best in occupied Germany.

Until the autumn of 1920 the German Road Service retained all the equipment lent them. After that six trucks from the A.F.G. proved sufficient until the end of 1921. The gasoline allowance of the A.F.G. had been reduced and the German Road Service could not procure sufficient funds for the operation of so many trucks; hence all but one truck were then turned back to the A.F.G.

The Interallied Road Commission, which functioned under the

Director Général des Communications et Ravitaillement aux Armées at Wiesbaden, did much good work in the co-ordination of road policies throughout the various zones. The German Road Administration was centralized in each area, and a single German authority was put in charge of the road service for the whole of the occupied territory. A system of strategic roads, called Inter-allied Routes, was determined upon, and it was agreed that each army should see to the maintenance of the portions of those routes that lay within its area. In 1920 an interesting study was made of the routes suitable for transportation of heavy ordnance material (up to 44 tons). Itineraries involving a nimimum of preparation in the reinforcement of bridges and culverts and in the widening of narrow streets through villages, were drawn up. To aid the Road Commission in its work, the French Road Service conducted experiments demonstrating the methods of reinforcing existing structures. Actual tests were made by passing heavy rollers and a 16-ton gun carriage over certain points.

As the appropriations made by the German Government were not adequate to maintain the roads at the standard required by the forces of occupation, the Interallied Road Commission decided that the various armies were empowered to order the German authorities to perform such additional work in their area as was required to keep the roads in proper condition. In the American zone the procedure adopted was to inform the German road delegate of the portions of road which reconnaissance had shown to be in unsatisfactory condition and to instruct him to prepare detailed estimates of the work necessary. Upon approval of the plans by the American Headquarters, the work was ordered to be performed.

In addition to the supervision of construction programs and the ordering of necessary repairs, the Officer in Charge of Public Utilities did much to assist the road officials in the matter of administration, obtaining deliveries of material and co-ordinating various municipal and public undertakings. For example, when on account of the heavy military traffic on a street in Coblenz it became necessary to pave it, that officer had the excavation for a new water main and the re-laying of the street railway tracks take place at the same time, thus necessitating but one tearing up of the street.

OFFICERS.

During the greater part of 1920 Captain F. Russel Lyons, Corps of Engineers, was in charge of public utilities and roads. He was succeeded by Captain Robert K. Munroe, C.E., who continued in that position until March 1922.

CHAPTER NO. 7

FOOD SUPPLY OF THE CIVIL POPULATION.

Introduction; Bread Supply; Meat Supply; Fats; Milk; Potato Supply; Other Foodstuffs; Conditions.

REFERENCE.

General . . . . . . A. M. G., Volume I, Chapter 8.

FOOD SUPPLY OF THE CIVIL POPULATION.

During 1920 the food situation in the occupied territories as well as in unoccupied Germany displayed a constant improvement. With the exception of the child relief work,[1] the Americans discontinued their auxiliary measures and the food supply of the civil population was left entirely in the hands of the civil population. All restrictions upon the purchase of German food by Americans were removed, and in that respect conditions became normal. The measures put into effect by the German Government during 1920 and 1921 are briefly described below.

Bread Supply. The compulsory rationing of breadstuffs was continued during 1920, except that peas, beans and buckwheat were released from control. In June 1921 the control was modified by the adoption of a new method of assessment. Every farmer who devoted more than one hectare (2.472 acres) to the growing of wheat, barley or oats was required to deliver to the German Food Office a quantity of grain which averaged about 900 lbs. per hectare. Three-fifths of the amount might be delivered in the form of oats. The grain thus secured was added to by purchases made by the Government in foreign countries as well as in Germany; the population, with the exception of those whose needs were served by their own production of grain, were supplied with one loaf (3.5 German lbs - about 3.9 English lbs.) of bread per individual per week. This bread was sold to those holding bread cards, at a price fixed by local councils. Additional bread could be purchased at discretion, but it was much more expensive as the flour was of a better grade than that used in the rationed bread.

[1] Volume II, Chapter 8.

All restrictions upon the purchase of such foodstuffs as Quaker oats, infant food, etc., had been removed by the end of 1920.

Meat Supply. The compulsory rationing of meat continued in effect during the greater part of 1920, but because of the exceptional conditions in the occupied territories the regulations were not strictly enforced there. In September the compulsory rationing was abolished, while certain temporary measures for conserving the meat supply were prescribed. At the end of 1921 the only remaining regulations related to professional dealing in cattle and to the conduct of cattle markets.

Fats. In June 1920 the restrictions concerning the rationing of margarine were removed, and its sale in the open market was authorized. The supplying of lard to the various districts was discontinued on December 1, 1920, an emergency reserve stock being retained by the Government for use in urgent cases. On April 30, 1921, the restrictions upon the sale of butter were removed, after which there remained no regulations concerning fats.

Milk. By the end of 1921 all restrictions regarding the home consumption of milk by producers had been removed. District administrations were still empowered to regulate the distribution of milk by certain rationing measures, and other regulations regarding the regular supply of milk to localities where shortages existed remained in effect. Producers were free to fix their own prices, but the Government reserved the right to provide for emergencies and the Oberpräsident, in his discretion, could fix prices should the necessity arise.

Potato Supply. All restrictions with regard to potatoes were abolished on September 15, 1920 by a decree of the Government. It had been presumed that the requirements of the population for winter potatoes had been provided for by the contracts made between the producers and the wholesale trade. But prices had risen by the time the potatoes were ready for delivery, the farmers refused to deliver them at the lower prices that had been agreed upon, and the wholesalers found it impossible to fulfill their contracts with the retailers. All barriers to the free circulation of potatoes were then raised, the only remaining proviso being that wholesale dealers in potatoes must be licensed.

Other Foodstuffs. By the end of 1921 there remained no restrictions upon other foodstuffs not mentioned above. As matters stood, all food control measures, except with regard to breadstuffs, had been abandoned.

CONDITIONS.

On January 1, 1922, the food situation in Germany was good. Prices were high and for that reason many of the poorer classes, including most of the salaried people, were unable to provide themselves with their full normal requirements. But the situation as a whole, as compared with previous years, was unquestionably good, and the food supply was no longer a serious problem even though some food had to be imported at high cost.

CHAPTER NO. 8

CHILD RELIEF IN THE AMERICAN AREA.

Conditions among Children in 1918; Hoover Commission; Distribution of Food; Method of Feeding; Results.

CHILD RELIEF IN THE AMERICAN AREA.

When the American troops entered Germany in December 1918 it was evident that many of the inhabitants of the occupied area, particularly the children under fourteen years of age, were undernourished. There was much malnutrition, rickets and tuberculosis among the children, many of whom were undersized and had weazened little faces. Children of this kind were noticeable in every city and town in the American area.

The soldiers, always fond of children, fed many of them; children gathered about a company kitchen, receiving food from the men, was a common sight. Officers and soldiers gave small amounts of food to the families with whom they were billeted, and naturally there were numerous cases of individual charity, but during the first year of the occupation there was no organized relief work.

In April 1920 Mr. Pierrepont B. Noyes, then American Representative with the High Commission, brought the conditions in the American area to the attention of the Hoover Commission in the United States, and recommended that a system of child relief work be instituted as rapidly as possible. The Hoover Commission immediately donated an initial fund of $50,000 for that purpose, and Mr. J. A. Dolan, of the American Department of the High Commission, was appointed director. He organized a system of distribution similar to that used in unoccupied Germany by the American Friends Service Committee. In June 1920 Mr. Noyes was succeeded as American Representative by General Allen; the latter and his Deputy, Colonel D. L. Stone, then took an active interest in the work.

The German authorities agreed to attend to all details incidental to the actual distribution of the food, and the work was gladly undertaken by all concerned. German physicians, under the supervision of the Kreis Representatives of the American Department, made a physical examination of 111,471 children between the ages of 6 and 12. The examination showed that, due almost exclusively to undernourishment, over 15% of the children were in very poor condition.

The food was shipped in bulk to Hamburg where it was stored by the American Friends Service Committee. The director of the relief work in the American area drew food from the Hamburg stocks by making periodical requisitions upon that Committee. Transportation expenses from the United States to Hamburg were paid by the Hoover Commission; those from Hamburg to Coblenz were paid by the German Government.

Actual feeding was begun on May 6, 1920, the program providing for the regular feeding of 10,000 undernourished children for a period of five months. It was expected that after the new harvest the food situation would have improved to such an extent that relief work, after the expiration of the five month period, would not be necessary. But conditions displayed only a slight improvement after the harvest, and it was decided that the work should be continued. Through the efforts of General Allen new appropriations were secured from the Hoover Commission.

The ration consisted of lard, cocoa, sugar, condensed milk, rice, flour, peas and beans, in such proportions as to yield approximately 667 calories per meal. The food was distributed in the form of one meal per day, Sundays excepted.

As the work progressed the children who had been receiving the food were subjected to a new physical examination. Those displaying sufficient improvement were replaced by others who had not yet participated, or else had been subjected subsequently to under-nourishment. This system resulted in a gradual decrease in the number being fed by the various kitchens.

The work reached its maximum extent in August 1920, when 80 kitchens, feeding more than 12,000 children per day, were in operation. A total of 23,283 children were fed in 1920 and 5400 in 1921. At the beginning of 1922 over 1800 children per day were being fed from 11 kitchens. The total number of meals served in the American area during 1920 was 1,397,000, while during 1921 the number was 324,650. The total cost to December 31, 1921, was approximately $100,000, or about 5.8 cents per meal. But for the hearty and thorough co-operation of the Germans it would have been impossible to have conducted the work at such small cost.

It was planned to continue the relief work in 1922 until such time as conditions should warrant its discontinuance.

CHAPTER NO. 9

FINANCE.

Agencies; Currencies Handled; Procurement of Funds; Disbursements; The Costs of Occupation; Computing the Costs of Occupation; Securing Partial Payments from Germany; Costs of American Occupation; Miscellaneous; Officers; Appendix.

REFERENCE:

General . . . . . A. M. G., Volume I, Chapter 12.

FINANCE.

AGENCIES.

Prior to July 1, 1920, all matters pertaining to disbursements of and accountability for public funds were conducted by the Disbursing Branch of the Quartermaster Corps. After that date, in accordance with Section 9-a of the National Defense Act, which created the Finance Department as a separate bureau of the War Department, the duties mentioned were handled by the Finance Officer, A.F.G. In August the duty of computing the costs of the American occupation was taken over, and in September the Auditor, A.F.G., came under the administrative control of the Finance Officer.

During 1920 and 1921 there were three Finance Officers in the A.F.G. - the central office at Coblenz and the branches at Andernach and Antwerp. The duties of the Coblenz and Andernach offices were largely of a similar nature; they made payments to military personnel, paid purchase and personal service vouchers for all services to the A.F.G., including German labor, and paid rentals to the various burgermeisters in the American area. After December 1, 1920, rentals for billets were paid by the German Government, which also took over the payment of German labor used by the A.F.G. after September 1, 1921.

In addition to making all disbursements for the Port of Antwerp, the Antwerp office settled all bills incurred through the return of members of Haller's army to the United States and the repatriation of the Czecho-Slovaks from Siberia,[1] agent officers

[1] Volume II, Chapter 23.

being sent for this purpose to Danzig and Trieste, respectively. The Antwerp branch was charged with making disbursements to settle the accounts of the Graves Registration Service at Antwerp, Bordeaux, St. Nazaire, Brest, Cherbourg, Dunkerque, Calais, London, etc. In June 1921 it sent an agent officer to Rome to pay the expenses arising because of the testing, acceptance and preparation for shipment of the ill-fated airship "Roma".

CURRENCIES HANDLED.

Until May 1, 1920, disbursements were made almost entirely in German marks, French francs and Belgian francs. Disbursements at Coblenz and Andernach were made largely in marks, the relative amount of French francs used decreasing as time went on; the Antwerp office handled Belgian francs for the most part.

After May 1, 1920, cash payments to military personnel and American civilian employees were made in marks or dollars at the option of the payee. Officers might take their pay in French francs, but by check only, or if they so desired, could take dollar checks on the U.S. Treasury; others were paid in cash.

Payment for labor and for supplies purchased in the American area were paid in marks, Belgian francs being used for similar purposes in Antwerp.

PROCUREMENT OF FUNDS.

The United States currency used by the A.F.G. was received from the following sources:

War Department - shipped by Army transports.
Sale of United States Treasury Checks.
Deposits of enlisted men.

Army Post Offices - sales of stamps and money orders.
Sales of German marks.
Exchanges of German marks - made for authorized persons only.
Quartermaster Corps - Sales Commissary funds.

The German currency was procured by the following means:

German Government - Armistice Funds.[1]
Enemy War Material - sales.
Government property - sales.
Subsistence stores - sales to German authorities for relief of civil population.
Provost Courts - fines.
Miscellaneous.[2]

French francs were mainly procured by invoice on the Financial Requisition Officer at Paris; they were placed to the credit of the disbursing officer of the Finance Office (Coblenz) with the Paris branch of the Guaranty Trust Co. of New York, and were drawn out by check. A small amount of French francs were acquired by exchange for dollars or marks. The rate of exchange was fixed every twenty days by the United States Treasury.

The Belgian francs were secured at first by the sale of United States Treasury checks. A large depository balance of Belgian francs - for Antwerp disbursements - had to be carried. Early in 1920, a heavy loss resulted from the depreciation of the Belgian

[1] A term used to denote German payments towards the costs of the American occupation; i.e., made under the Armistice Agreement and not under the Treaty of Versailles.

[2] Receipts from Q.M. laundry; Sales of confiscated liquor; A.P.O. 927 - sales of stamps and money orders (until spring of 1920 when it was required that such sales be made for U.S. currency only); railway and berth fares on special sleepers between Coblenz and Antwerp; fines collected by Rhine patrol boats; rent of transient rooms in Coblenzerhof and Monopole Hotels; War Department - for Educational and Vocational Training; rentals from Motor Organizations, Limited; transfers from officers.

franc. During the calendar year 1920 the losses amounted to $559,309.96; these were offset in part by a gain of $178,146.23. During 1921 the loss was $4,185.02, while the gain was $15,906.95. After May 1921 the Belgian francs needed were secured through the sale of German marks - another method of reducing the amount of the depreciating German currency on hand. Up to February, 1921, the rate of exchange was fixed every twenty days by the United States Treasury, and after that the commercial rate was used.

DISBURSEMENTS.

The total disbursements in United States currency of the Coblenz, Andernach and Antwerp offices during 1920 and 1921 were as follows:

Period	Disbursements	Monthly Average.
1920	$14,738,465.42	$1,228,205.45
1921	10,440,387.39	870,032.28
Total	$25,178,852.81	$1,049,118.87

During the same period over 650,000,000 German marks were disbursed as follows:

1920	339,471,151.44
1921	314,313,704.45
Total	653,784,855.89

THE COSTS OF OCCUPATION.

One of the most interesting questions that arose in connection with the American occupation had to do with the cost of the maintenance of the American troops (generally called the costs of occupation) and the efforts to collect the large sums due.

It may be said in passing that the A.F.G. continued to be subject to the embarassment cuased by the decision of the Assistant Comptroller of the Treasury - that all monies received from the German Government had to be covered into the Treasury, and could only be disbursed therefrom in accordance with Congressional appropriations.[1] Had he decided otherwise, it was believed by the American authorities of occupation, not only would their financial problems have been less intricate, but also it would have been possible to have secured from the German Government far more funds with which to defray the upkeep of the troops.

It will be remembered that Clause IX of the Armistice Agreement provided that the upkeep of the troops of occupation in the Rhineland should be charged to the German Government. The Treaty of Versailles made ample provisions for the collection of the costs of occupation. Article 235 provided that out of the 20,000,000,000 gold marks to be paid by May 1921, the expenses of the armies of occupation should first be met. Article 248 provided that subject to such exceptions as the Reparations Commission might approve, the cost of reparation and all other costs arising under the Treaty or supplementary treaties or under arrangements concluded under the Armistice or its extensions, should be a first charge against the assets and revenues of Germany. Article 249 stated that Germany should pay the total costs of all the armies in occupied German territory from the date of the signature of the Armistice of November 11, 1918. Article 251 established an order of priority of the charges provided for in Article 248; first were to be the costs of occupation during the Armistice and its extensions, and second were to be the costs after the coming into force of the

1 A. M. G., Volume I, Page 284.

Treaty; reparations were to be third and other obligations "incumbent upon Germany" fourth.

The Treaty of Versailles and the Rhineland Agreement were signed on June 28, 1919. During the period from that date until they came into force on January 10, 1920, the upkeep of all the armies of occupation was payable and susceptible of collection under the Armistice Agreement and extensions thereof.

The Position of the United States.

On December 24, 1919, Mr. Henry A. White, the American representative made the following statement before the Supreme Council:

> Without desiring to prevoke discussion but in order to leave no room for any misunderstanding, the United States Government in desirous to state its position of record as follows:
>
> 1. The Annex to Article 88 of the German Treaty of Peace provides that "the cost of the Army of Occupation" in Upper Silesia will be a charge on the area. Other articles by similar language provide for the payment of the expenses of the Armies of Occupation in the Occupied districts. In like manner, Article 235 of the German Peace Treaty provides that the expenses of the Armies of Occupation subsequent to November 11, 1918, shall be paid out of the first sums received out of reparation.
>
> 2. United States, therefore, has always understood that it is entitled to be reimbursed the exact cost of maintaining all its military forces in German territory or in the several plebiscite districts since November 11, 1918, until finally withdrawn.
>
> 3. While there have been many conferences on this subject and while the matter has been previously discussed in the Supreme Council, no representative of the United States Government has ever agreed nor has had any authority whatever to agree to any other position or view.
>
> 4. If it be considered that the cost of the United States forces in any of the occupied territory or plebiscite districts three thousand miles from their base is greater, or will be greater than is desirable, it will be recalled that Associated Powers have repeatedly requested the United States

> to consent to their use for this purpose. If the Associated Powers deem it to their interest to reduce the cost of the Armies of Occupation, it would appear to be preferable to request the United States Government to decrease its military representation rather than to resort to financial expedience, the legality of which might be questioned.

After the ratification of the Treaty the payment of the costs of occupation was to be secured under the provisions of the Articles already mentioned. It became increasingly evident, however, that the Treaty would not be ratified by the United States and that an embarassing situation would result. The Committee on Organization of the Reparations Commission was holding frequent meetings during which there was considerable discussion relative to the costs of occupation and more particularly to the upkeep of the American forces. At the 29th meeting on January 7, 1920, the American delegation submitted the following note:

> The Supreme Council on December 9, 1919, having referred back to the Organization Committee of the Reparation Commission, for further examination, the question of principle involved in the payment of the cost of the Armies of Occupation, the American Delegation regrets that it cannot recede from the former position taken by the United States as stated by the Honorable Henry A. White, of the American Peace Delegation, at the Meeting of the Supreme Council of December 24, 1919.
>
> Mr. Rathbone, American Delegate on C.R.R.C. stated at this Meeting that the general point of view of the United States was expressed in the foregoing document and that this point of view was final.

The Treaty of Versailles and the Rhineland Agreement came into force on January 10, 1920, except in so far as the United States was concerned. The position of the United States was defined in the following note from the American Ambassador to France to M. Clemenceau, President of the Peace Conference, under date of January 12, 1920:

My dear Mr. President:

Referring to the question of the jurisdiction of the Rhineland Commission, which was discussed at the Meeting of the Council on December 30th last, you will undoubtedly recall that the Council decided that as soon as the Treaty of Versailles went into force. the jurisdiction of the Rhineland High Commission would extend over all the territories of the left bank of the Rhine, including those occupied by the American forces. At that time it was agreed that I would inform my Government of the action of the Council and request an expression of its views. I have the honor to inform your Excellency that I have today received the following from my Government, which I am directed to communicate to the Council:

"The Government of the United States cannot acquiesce in decision of the Council that as soon as the Treaty of Peace goes into force between the Allied Governments and Germany, the jurisdiction of the Interallied Rhineland High Commission to be created under the Rhineland Agreement signed June 28, 1919, will extend over all the territories on left bank of the Rhine, including those occupied by the American troops.

The views of the Government of the United States are that while upon the Treaty of Peace coming into force, the Armistice, except for provisions specifically continued in force, will terminate as between Germany and the Powers ratifying the Treaty, it nevertheless continues in force as to the United States.

Until the Treaty of Peace with Germany and the Rhineland Agreement are ratified by the United States, the belligerent rights of the Government of the United States in the Rhenish territory occupied by its forces under the Armistice can in no way be effected or limited by Agreement between the Allied Governments and Germany not ratified by and therefore not binding upon the United States."

The Government of the United States, therefore, cannot agree that the jurisdiction of the Rhineland Commission to be created upon the coming into force of the Treaty will extend at this stage over American troops or the Rhenish territory occupied by them, but nevertheless it will be disposed, should the Allied Governments so desire, to have an American representative unofficially present at the meetings of the Commission to be created,

> and insofar as practicable, cause to be put into effect concurrently in the American zone of occupation, measures similar to those to be adopted by the Commission for the portions of the Rhenish territory occupied by the Allied Governments, provided the measures do not conflict or exceed the terms of the Armistice.
>
> I am, etc ..... (Signed) Hugh C. Wallace.

On January 14, 1920, the German Government was informed of the position of the United States in a memorandum for Mr. J. C. Grew, Counsellor attached to the American Embassy at Paris, to Baron von Lersner of the German Peace Delegation -

> The Government of the United States regards the Armistice between the United States and Germany as remaining in full force and effective, notwithstanding the deposit of ratifications of the Treaty of Peace.
>
> An official notification to this effect is being sent by the Government of the United States to the Swiss Minister in Washington for transmission to the German Government.

Through its failure to ratify the Treaty of Versailles and the Rhineland Agreement, the United States was, in effect, left to make its own arrangements for collecting the costs of the American occupation. On the other hand the ratifying Powers, under the terms of the Treaty, could act through the Reparations Commission and would probably demand all that Germany could pay. The United States could demand direct payment from Germany, but this action would raise the question of priority of payment - whether preference should be given to the payment of the costs of the American occupation or to all reparations demanded by the Allies.

While the United States was not officially represented on the Reparations Commission, Mr. R. W. Boyden attended all meetings in the capacity of United States Unofficial Delegate. The Commission was authorized by the ratifying powers to interpret and execute Part IX of the Treaty of Versailles, and on April 19, 1920, accepted the mandate in the following decision.

> It was decided that the Reparation Commission should accept the Mandate which had been given it by the Governments of the principal Allied and Associated Powers (with the exception of Japan and Belgium), to interpret and execute the clauses of Part IX for which the Commission was not already competent, under the specific terms of the Treaty.

The question of the costs of the American occupation constantly entered into the deliberations of the Reparations Commission and the position of the United States was frequently and consistently defined. At the 69th meeting on July 26, 1920, the following action was taken:

> It was decided that the delegates should send a separate letter to their respective Governments, drawing the attention of the latter to the amount of expenses necessitated hitherto by the Armies of Occupation, and informing them that from the point of view of reparation, it would be desirable to reduce such expenses.
>
> On the proposal of Mr. Boyden, the necessary modifications would be made in the letter to be sent to the American Government, in order to take into account the fact that the position of the American Army of Occupation is at present governed by the terms of the Armistice and other agreements and not by the Treaty.

Again, at the 73rd meeting on August 5, 1920, Mr. Boyden addressed the Reparations Commission as follows:

> In view of the arrangements between the Powers concerned as to the priority to be afforded to their advances under the terms of the Spa Protocol, the United States Unofficial Delegate makes no protest or reserve, merely pointing out that these decisions, to which the United States is not a party, do not effect the position of the United States. With respect to the costs of the United States Army of Occupation, he adds that

the United States obviously understands and expects that it will be reimbursed in cash for the actual cost, and that it will be notified at once if its army is not wanted on these terms. In this connection, he refers to the letter of the American Delegation to C.O.R.C., dated November 28th, 1919, and its accompanying memorandum.

On January 11, 1921, a cablegram from the State Department to Mr. Boyden clearly defined the position of the United States, at the same time authorizing the Reparations Commission to make collections in its behalf -

2. Your 7. Department considers Rhineland agreement of June 28, entered into in pursuance of Article 432 as supplemental to Treaty. This agreement submitted to the Senate prior to Treaty coming into force as to other Powers on January 10, and has not been ratified. This Government pending ratification of Treaty and agreement, not bound thereby. Agreement cannot be considered an Armistice Agreement since it regulates occupation after conclusion of peace.

3. Under Armistice which is still in force as to United States, this Government clearly has right to collect cost of occupation direct from Germany. In certain eventualities it might become desirable to collect directly. Department, however, has interposed no objection to apparent desire of Reparations Commission to make collections on this Govenment's behalf.

4. This Government cannot, of course, at this time, claim benefits of Treaty as a ratifying Power. It has inchoate rights in assets of Germany disposed of under Treaty, which cannot be defeated by other Powers. Furthermore, under Article 248, Germany agrees that cost of reparations and all other costs arising under the Treaty or "other arrangements concluded between Germany and the Allied and Associated Powers during the Armistice or its extensions" shall be a first charge on all the assets and revenues of the German Empire or its constituent States. By Article 249, Germany agrees to pay total costs of "all armies of the Allied and Associated Governments in occupied territory from the date of the signature of the Armistice of November 11," Article 251 provided for such costs (a) during the Armistice and its extensions, (b) after the coming into force of the Treaty, shall have priority over other charges. This priority charge on her assets is a definite obligation on Germany which is now bound by Treaty and cannot be denied

by her or other principal Allied Powers also bound by Treaty. Since it covers the costs of "all" Allied and Associated armies in, and it clearly applies to the United States, which were placed in occupation under the Armistice and in pursuance of understanding with principal Allied Powers. (In this connection see statement of Mr. White, November 24, 1919, M.D. 99, Annex F, that Allied forces repeatedly requested United States Government to use American troops for purpose of occupation). Therefore, irrespective of ratification of Treaty by the United States, this Government is entitled to the benefits of this lien and priority for its army costs. This Government cannot admit that either Allied Governments or Reparations Commission can dispose of assets thus pledged without taking into consideration interests and rights of the United States in respect thereto. Since Germany is definitely bound by lien created for the costs of all armies of occupation, and Powers ratifying Treaty are also bound to recognize this lien, no further agreement would seem to be necessary, and for foregoing reasons, Department does not consider that payments of American costs by Commission would be, as indicated in your paragraph fifth, a waiver of security and priority of other Powers.

It was expected that this stand would result in equitable consideration and treatment of the costs of the American occupation, despite the failure of the United States to ratify the Treaty. But the Reparations Commission apparently decided that such action was beyond its authority, unless specifically so instructed by the ratifying Powers. Up to December 31, 1921, the claims of the United States had not been considered in the distribution of the reparations payments made by Germany. In fact the United States appeared to be in the position of an unsecured creditor of Germany, all of whose property was already mortgaged to other Powers.

The ratification on November 11, 1921, of a Treaty of Peace (the Treaty of Berlin) between the United States and Germany did not appear to have altered matters, although the United States had reserved to itself all rights acquired under the Armistice Agreement and those stipulated in the Treaty of Versailles.[1] The Treaty of Berlin raised an interesting question: Could the United States, a non-ratifying party to the Treaty of Versailles, in a subsequent treaty with a ratifying party (Germany) legally reserve to itself any rights or claim any benefits under the terms of the former Treaty, without the consent of all ratifying parties to that Treaty?

The consistent stand of the United States was as follows:

> The United States was entitled to receive in cash the upkeep of its troops of occupation in the Rhineland, regardless of the amount. The cost of this upkeep, together with similar claims of the Allies, was a first mortgage against the assets of the German Government. No Allied Power might make with Germany any treaty arrangements which would prevent the United States from recovering the costs of the American occupation on the same footing as the Allied Powers which had ratified the Treaty of Versailles.

Computing the Costs of Occupation.[2]

Until the ratification of the Treaty of Peace between the United States and Germany on November 11, 1921, the costs of the American occupation had been charged under that portion of clause IX of the Armistice Agreement which stated that

> The upkeep of the troops of occupation in the Rhineland (except Alsace-Lorraine) shall be charged to the German Government.

In the Treaty mentioned above, the United States reserved to itself all rights which it would have acquired had it ratified the Treaty of Versailles, and after November 11, 1921, the costs of the American occupation were computed under the latter treaty. Actually, there was practically no change in the method of

[1] Volume I, Chapter 1.
[2] See Appendix to this chapter.

computation or in the items of upkeep included, as the system used by the A.F.G. was in accord with the views of the Reparations Commission and had been approved by that body. Article 249 of the Treaty of Versailles reads:

> There shall be paid by the German Government the total cost of all armies of the Allied and Associated Governments in occupied German territory from the date of the signature of the Armistice of November 11, 1918, including the keep of men and beasts, lodging and billeting, pay and allowances, salaries and wages, bedding, heating, lighting, clothing, equipment, harness and saddlery, armament and rolling stock, air services, treatment of sick and wounded, veterinary and remount services, transport service of all sorts (such as by rail, sea or river, motor lorries), communications and correspondence, and in general the cost of all administrative or technical services the working of which is necessary for the training of troops and for keeping their numbers up to strength and preserving their military efficiency.
>
> The cost of such liabilities under the above heads so far as they relate to purchases or requisitions by the Allied and Associated Governments in the occupied territories shall be paid by the German Government to the Allied and Associated Governments in marks at the current or agreed rate of exchange. All others of the above costs shall be paid in gold marks.

Detailed instructions for computing the costs of the above items, including depreciation, were drawn up. It was required that the charges to Germany "must be limited to strict military necessities essential to the Army of Occupation as a military body in an enemy country; such as to be justified on this ground if they should be reviewed by an International Tribunal. They must not be of a class which could be considered merely desirable or convenient."

The Auditor, A.F.G., who was under the supervision and control of the Finance Officer, was charged with the duty of computing the costs of the American occupation. Each quarter he received from the various Services detailed statements of their respective expenses of upkeep. The Auditor then compiled a consolidated quarterly report, copies of which were forwarded to the American Unofficial Delegate with the Reparations Commission; the latter official submitted the reports to the Reparations Commission which presented them to the German Government. Other copies were forwarded to the Adjutant General of the Army and to the Chief of Finance.

A recapitulation of the costs of the entire period of occupation was carried forward each quarter; this paper showed the total charges to date, the total amount of paper marks credited to Germany, and the amount due the United States.

The costs of the American Department of the Interallied Rhineland High Commission were computed separately from the costs of the A.F.G. and separate reports thereof were rendered. The quarterly report, however, carried a grand total of costs, including the American Department, setting forth the entire charges, credits and balance, for the period from the beginning of the occupation to date.

Securing Partial Payments from Germany.

As the United States, because of the actions of the Reparations Commission, had not shared in the reparations payments made by Germany, there was but one means of reducing, even in a small way, the amount due for the upkeep of the American troops in the Rhineland. This was to requisition from Germany such amounts of paper marks as could be used for disbursements in Germany.

These amounts were limited by reason of several factors. Primarily, most of the supplies used by the American troops were purchased in the United States and shipped to occupied Germany. Large amounts of the property and supplies rendered surplus by the reduction of the forces were sold for German marks; until November 1920, because of the impracticability of converting them into United States currency, these marks were used for local disbursements, in lieu of those that otherwise would have been obtained by requisition upon the German Government.

Again there was the question of the method of payment of officers and soldiers - in dollars or in marks. Prior to May 1, 1920, officers, soldiers and civilian employees received their pay in German marks at rates of exchange fixed monthly by the Treasury Department. Almost invariably the rate was unfavorable to the payee, as the local banks would pay more for the dollar than was authorized by the official rate. This caused considerable dissatisfaction, and authority was obtained from the War Department to pay the A.F.G. in dollars or marks, at the option of the payee, effective from May 1, 1920. Because of the great depreciation of the German mark (it went as low as 330 to the dollar in November 1921), officers and men usually elected to receive their pay in United States currency. But occasionally the official rate of exchange authorized more marks to the dollar than did the local banks; at such times the troops would demand their pay in marks. This factor required rather a large reserve of marks to be carried by the Finance Officer, as it was necessary to notify the German Government fifteen days in advance of the date marks would be needed. If the official rate of exchange happened to be unfavorable, the troops would draw their pay in dollars, and the marks on hand would have to be carried over to another accounting period. While there were fluctuations in the value of the mark, the depreciation grew from month to month, with the result that a loss in the value of the marks on hand had to be written off. These losses amounted to several millions of dollars. To avoid the losses and to increase disbursements, the Commanding General directed that for disbursing purposes a daily rate of exchange be adopted. Arrangements were made with the German Government whereby marks were disbursed at a daily rate of exchange fixed by that Government; Germany then received credit on the account of the upkeep of the A.F.G. for the exact value of the marks disbursed at the rates of exchange thus fixed. This system was put into effect on October 26, 1921, after which there were no losses in the marks received from Germany.

In the spring of 1919, through sales of surplus property and supplies and of enemy war material, a large amount of marks had been accumulated by the American Army of Occupation. The Germany currency was rapidly depreciating, and at the time it was impossible to convert the marks into United States currency; therefore, between May 31, 1919 and February 27, 1920, no funds were requisitioned from the German Government.[1]

On January 1, 1920, the Armistice Funds remaining on hand amounted to over 193,000,000 marks. These were disbursed during

[1] A. M. B., Volume I, Page 288. Funds had been secured by requisition upon the German Government (National Property Administration) which deposited them in the Coblenz branch of the Reichsbank.

January and February, in payment of troops and for supplies and services obtained locally.

The German marks secured through sales of supplies, Provost Court fines, etc., continued to accumulate during 1920 and 1921. In the autumn of 1920 it was decided to obtain from the German Government all the marks that could be disbursed in payment of the upkeep of the A.F.G. and to dispose of those received from other sources by selling them for United States currency whenever possible.[1]

Costs of American Occupations.

The total costs, chargeable to Germany, of the American Army of Occupation from the beginning of the occupation, December 1, 1918, to December 31, 1921, were as follows:

Period	Costs	Less value of paper marks received from German Gov't.	Balance due United States
1918-19	$242,319,147.10	$29,610,086.60	$212,709,060.50
1920	24,866,686.25	3,752,093.69	21,114,592.56
1921	20,446,873.52	2,635,766.31	17,811,107.21
Totals	$287,632,706.87	$35,997,946.60	$251,634,760.27

It will be noted that the major portions of the costs accumulated during the first thirteen months of the occupation. This was particularly true during the first six months when American troops in the Rhineland numbered approximately 250,000 men. When the Treaty of Versailles became effective there were

[1] Up to December 31, 1921, over 137,000,000 marks were sold under this policy. They were sold to local banks at the official rate of exchange fixed by the Treasury Department, or at a better rate when obtainable. A gain of $58,317.54 on these Sales was taken up by the Finance Officer.

due the United States more than $212,000,000, or about 88% of the costs to date. From January 1, 1920, to December 31, 1921, the costs of occupation were less than $45,500,000, of which approximately 86% remained unpaid. Of the total costs of the American occupation, about 87½% remained unpaid at the end of 1921.

MISCELLANEOUS.

Receipts and Disposals off German Marks.

The following is a statement of German marks received from all sources and the disposition made of them during 1920 and 1921:

RECEIPTS

Period	Rec'd from Sales, etc.	Armistice Funds	Totals.
Balance on hand Dec. 31, 1919			193,373,826.61
1920	81,395,051.92	245,000,000.00	362,395,051.92
1921	38,307,840.00	257,041,517.40	295,349,357.40
Totals	119,702,891.92	502,041,517.40	851,118,235.93

DISBURSEMENTS, EXCHANGES AND SALES.

1920	Disbursed	339,471,151.44
1921	Disbursed	314,313,704.45
Nov. 18, 1920 to Dec. 31, 1921	Marks sold for US Currency.	137,117,229.89
1920	Marks exchanged for authorized persons	13,194,801.30
1921	Marks exchanged for authorized persons	4,843,557.07
Total marks disbursed, sold and exchanged, 1920-1921		808,940,444.15
Balance marks on hand December 31, 1921		6,177,791.78

Prior to January 1, 1921, purchases at the Sales Commissary

could be paid for in marks. From this source 20,743,088.19 marks accumulated; these were in addition to the amount of receipts shown above. Approximately 17,500,000 of these marks were sold, and the remainder were disbursed locally.

The German marks received during 1920 and 1921 from sources other than the German Government (Armistice Funds) and the Sales Commissary were as follows:

Sales of enemy war material	8,788.50
Sales of Surplus property and supplies	71,067,415.19
Sales of subsistence for relief purposes	323,080.78
Fines by Provost Courts	1,034,561.00
Miscellaneous	36,877,282.16
Exchanges	10,391,764.29
Total	119,702,891.92

From the beginning of the occupation until December 31, 1921, the marks received from all sources other than the German Government (Armistice Funds) amounted to 293,365,888.76; the total (including 1920 and 1921) received from these sources amounted to 413,068,780.68 marks. The total from all sources, including Armistice Funds, for the period December 1, 1918, to December 31, 1921, was as follows:

Armistice Funds	810,041,517.40
Other sources	413,068,780.68
Total	1,223,110,298.08

A striking illustration of the depreciation of the German currency is shown by the statement of the marks received from the German Government and their equivalents in dollars, during the calendar years 1918 to 1921, inclusive --

Year	Marks Received	Dollar Equivalent
1918	54,000,000.00	6,770,931.21
1919	254,000,000.00	22,839,155.40
1920	245,000,000.00	3,752,093.69
1921	257,041,517.40	2,635,766.31
Totals	810,041,517.40	35,997,946.61

A tabulation of the high and low values (expressed in the number of marks that could be bought for one dollar) of the mark during each of the calendar years mentioned, will complete the picture, especially when it is remembered that the pre-war value of the mark was about 23 cents or that one dollar was worth about 4.4 marks --

Year	High (marks for $1)	Low (marks for $1)
1918 (Dec. only)	7.78	7.78
1919 (to May 31 only)	8.13	13.44
1920	36.36	103.21
1921	59.52	303.03

The rates of exchange shown above were the official rates at the time the funds were received from the German Government. In 1920 the mark depreciated as low as 120 to the dollar, and in 1921 its lowest point was 330 to the dollar.

OFFICERS.

Lieutenant Colonel R.A. Sturges, Finance Department, served as Finance Officer, A.F.G. throughout 1920 and 1921. During the same period the following officers served under him:

Coblenz

Major E. E. Lockard, F.D.
" E. N. Enders, F.D. (Andernach)
" H. G. Rice, F.D.
" J. H. Walton, Q.M.C.
Captain O. A. Greager, Q.M.C. (Andernach)
1st. Lieut. J. H. McFall, F.D. (Auditor, A.F.G.)

Antwerp.

Major G.R. Doane, F.D.

- - - - - - - - - -

APPENDIX

COMPUTATION OF THE COSTS OF OCCUPATION.

The Quartermaster (Statistical Division) was charged with the important duty of computing the cost of the upkeep of the A.F.G. under the various subjects listed below. These figures represented approximately 90% of the total cost of the A.F.G. The data therefor was secured monthly from the various offices and a quarterly compilation thereof was submitted to the Auditor, A.F.G. The various expenses were consolidated under the following headings:

Officers

Soldiers

Animals

General Expenses

Debits

Credits

German Labor

Supplies and Services furnished in kind by the German Government.

Officers. Under this heading were consolidated the amounts arising from the pay and allowances of the officers of the A.F.G.

Soldiers. This included the pay and allowances of the soldiers, and the cost of the food, clothing, personal equipment, bedding, heat, light and water furnished them.

Animals. The cost of remounts and of forage, equipment, stable material, etc., were included in this item.

General Expenses. Coming under this heading were the costs of railway, barge, horse and motor transportation, and of gasoline and oil, Quartermaster repairs and technical supplies, administration and training, field clerks, warrant officers, United States Civil Service Clerks, Allied and German employees, rentals and billets, and telegraph, telephone and postal service.

Debits. Included in this item were all claims for damage to or loss of American property (when such was incidental to the occupation), including properly authenticated surveys of property damaged by fair wear and tear.

Credits. The proceeds of sales of salvaged property, which had been originally charged to Germany at the full cost, were credited to that country and deducted from the costs of Occupation.

German Labor. Until July 31, 1921, German employees were paid from Quartermaster funds. After that date, by arrangements with the Reichsvermögensamt (National Property Administration), all German employees were paid by the German Government in accordance with wage scales authorized by the American Headquarters and published from time to time. The American authorities retained the right to employ, discharge and control all such employees. This arrangement resulted in a saving of approximately $9,000 per month in funds expended by the Quartermaster for labor.

Supplies and Services furnished in kind by the German Government. From the beginning of the occupation until early in 1920, all supplies and services obtained by requisition upon the Germans were paid for by the American authorities, the necessary funds being secured by requisition upon the German Government. Thereafter most of the supplies and services were furnished in kind by the German Government, which also made the necessary payment therefor. If any supplies secured under this arrangement were sold or otherwise disposed of except for the A.F.G., the value was credited to Germany and set off against the cost of occupation. An accurate record of all transactions under this heading was kept and included in the periodical reports made to the American Unofficial Representative with the Reparations Commission.

OFFICERS

Major C. L. Gilbert, Field Artillery, was in charge of the Statistical Division of the Quartermaster's office throught 1920 and 1921.

CHAPTER NO. 10

RENTS, REQUISITIONS AND CLAIMS SERVICE.

Inauguration; Organization; Rents; Requisitons; Claims; Claims under Act of April 18, 1918; Luxembourg Agreement; Officers; Miscellaneous - General Orders No. 104, Instructions for the R.R.& C. Service, Sample Forms.

REFERENCE.

General . . . . . . A. M. G., Volume I, Chapter 12.

RENTS, REQUISITIONS AND CLAIMS SERVICE.[1]

INAUGURATION

The march of the Third Army to the Rhine made necessary the organization of a provisional service to secure accomodations, such as quarters, camp sites and stables, for the advancing troops. This service was called the Rents, Requisitions and Claims Service, becoming generally known as the R. R. and C. The various units detailed officers as Town Majors; with the necessary clerical personnel these officers preceded the troops by approximately twenty-four hours, located and listed the available accomodations, and assigned them for the use of smaller units and individuals. The Town Majors were directed to keep accurate records of the billets and other accomodations used and to secure payment therefor through the Division Quartermaster. After the units of the Third Army had arrived at their stations in the American Occupied Territory, the R. R. and C. was given a more definite organization. Each division appointed its own R. R. and C. Officer and placed under him the Town Majors of the garrisoned towns in the divisional area.

During the first weeks of the occupation it was the general opinion that whatever might be considered necessary for the troops could be obtained by direct requisition upon the Germans, and that payment therefor would be made at some indefinite date in the future. In many cases articles or property were obtained by requisitions that were entirely unauthorized, and much confusion resulted. The activities of the R. R. and C. personnel were then confined to the requisitioning of the necessary real property, the

[1] A. M. G., Volume I, Chapter 12.

preparation and payment of billeting accounts, and the reception and investigation of claims submitted by the German inhabitants.

In October 1919 a central R. R. and C. office was established in Coblenz. All divisional and other pertaining records were sent to that office. It developed that there remained a great many unsettled claims and billeting accounts and a large number of unpaid rentals for requisitioned real property. The records were then consolidated, vouchers for the unpaid accounts were made, and the amounts due were paid to the burgermeisters concerned by the Disbursing Quartermaster of the A. F. G.

ORGANIZATION

The central office of the R. R. and C. Service consisted on January 1, 1920, of the R. R. and C. Officer, A. F. G., an assistant R. R. and C. Officer, and two Investigating Officers for the "Army Appraisal Board".[1] The field organization comprised Zone Majors in the Kreise of Coblenz (Land and Stadt), Mayen, Cochem, Adenau, Ahrweiler, Neuwied and Montabaur, with Town Majors in all garrisoned towns.

All requisitions and billeting accounts were forwarded through the Zone Majors to the central office for final action; the latter office prepared the vouchers and secured payment thereon. Claims submitted by the inhabitants were investigated first by the Town Majors who forwarded them, together with their recommendations, to the Zone Majors. After further investigation the Zone Majors forwarded the claims and their recommendations thereon to the central office. If the final investigation so indicated, the claims

[1] See G.O. No. 15, A.F.G., 1919, quoted further on under "Rents".

were vouchered and forwarded to the Officer in Charge of Civil Affairs for payment by the German Government as a part of the costs of occupation. Careful records were kept of all claims, whether or not approved.

RENTS

During the existence of the Third Army a board of officers was appointed in each divisional area to determine equitable rentals for requisitioned property. Values, taxes and costs of maintenance were considered in an attempt to fix rentals at amounts similar to those paid by the German inhabitants. The rentals finally established varied greatly, but no attempt to fix definite rentals for the whole American Area was made. On the other hand, billeting lists (rentals of single billets) were paid at rates that were the same throughout the American Zone. Rentals for the A. F. G. were fixed by Section II of General Orders No. 15, issued on September 9, 1919, which read as follows:

> 1. A Board of Officers is appointed to meet at these Headquarters, from time to time, to determine what constitutes a just rental price for each of the various buildings, premises, establishments, utilities, etc., requisitioned within area occupied, by authority of these Headquarters, as necessary for the needs of the American Forces in Germany.
>
> Such price having been duly determined by the Board and approved by these Headquarters shall govern payments made for the use of such buildings, premises, establishments, utilities, etc., by the American Forces in Germany.
>
> 2. In determining the rental to be paid in any case, the following factors will, in general, be considered by the Board:
>
> (a) VALUE OF INVESTMENT.
> (1) Value of Buildings.
> (2) Value of Grounds.
> (3) Value of Equipment.
> (4) Total.
>
> (b) ITEMS OF EXPENSE.
> (1) Depreciation of equipment.

(2) Repairs and maintenance.
(3) Heat, Light, Water.
(4) Insurance.
(5) Taxes.
(6) Payroll of necessary working force of employees, including janitor service, char-women, engineers, mechanics, elevator service, etc., necessary for the proper care and maintenance of buildings and grounds, facilities, etc., concerned.
The cost of services, supplies, kitchen and table ware or wages of employees engaged directly in the preparation and serving of food in restaurants and hotels taken over on requisition, will not be included in determining any cost of operation and maintenance or any payroll of employees.
(7) Total.

(c) TOTAL EXPENSES FOR THE YEAR.
(1) The relation between the totals of (b) and (a) above, should be determined on a percentage basis. This should not exceed 4% of (a) as a normal figure.
(2) That figure which would be regarded in this country as a reasonable and fair return on (a) above, should also be determined. It should not exceed 6% of (a) as a normal figure.
(3) The figure representing the just rental price, above referred to, should not exceed 10% of (a), as a normal figure.

(d) MONTHLY RENT DESIRED BY OWNER OR LESSOR.

3. DETAIL FOR THE BOARD.
Chief Quartermaster, American Forces in Germany.
Chief Disbursing Officer, American Forces in Germany.
Chief R.R.& C. Officer, American Forces in Germany.
Town Major, Coblenz, Germany.

4. A Board of Officers will be convened for purposes analagous to those mentioned above.
The Board so convened in the American Forces in Germany, will handle and bring to completion all matters of the above nature that properly lie within its jurisdiction.

This board was not given a particular title but it became generally known as the "Army Appraisal Board" and will be so designated in this chapter.

The German authorities requested that they be allowed a voice in the fixing of rentals for requisitioned real property and single billets, as such matters were obviously of much importance to the German inhabitants. This request was granted, and Par. 6, Special Orders No. 277, was issued on October 5, 1920 -

> 6. A Board consisting of the following named officers to be designated as "The Billeting Board, American Forces in Germany", is convened at these Headquarters, to meet at the call of the President for the purpose of taking up and making recommendations of such matters as may be brought before it:
>
> DETAIL FOR THE BOARD
>
> CHIEF QUARTERMASTER, A.F.G., President,
> CHIEF FINANCE OFFICER, A.F.G.,
> CHIEF R.R.& C. OFFICER, A.F.G.,
> TOWN MAJOR, City of Coblenz, Recorder.
>
> The following named German members are designated as members of the Board, and will sit with it at all of its meetings:
>
> DR. GAEDE, Regierungsrat, Office of the Oberprasident,
> DR. BINHOLD, Alderman, City of Coblenz.
>
> The Board will have at least one meeting every month, and as many more as deemed necessary by the President. The proceedings of the Board will recommendations on matters taken up will be forwarded to the Adjutant General, A.F.G.

One German member represented the Oberprasident of the Rhine Province and the other the Obergurgermeister of Coblenz. Two more German members were designated later, making four in all; of the new members, one was an additional representative of the Oberburgermeister of Coblenz, while the other was a member of the Reichsvermogensverwaltung (National Property Administration) under which a German Billeting Commission functioned. The duties of the Billeting Board were advisory. All matters concerning billets, rentals and requisitioned real property and equipment were first considered by the Billeting Board; the latter submitted its recommendations to the Army Appraisal Board which made the necessary decisions.

On December 1, 1920, the R. R. and C. Service discontinued the practice of securing real property be direct requisitions upon burgermeisters. The determination and fixing of rentals and the payment of "Billeting Distribution Lists" by the Army Appraisal Board was also discontinued. All of these functions were transferred to the German Billeting Commission, which had been organized for that purpose.

The German Billeting Commission consisted of four members, all of whom were well versed in financial matters and the values of real property. It was required that at least two of the members should be permanent residents of the American Area and owners of substantial amounts of real property therein. The President, who was also an official of the Reichsvermogensverwaltung appointed the other members whose assumption of office was subject to the approval of the American Headquarters.

The German Billeting Commission had the following functions:

> The requisitioning of real property, with the necessary equipment and operating personnel, when so directed by the R. R. and C. Officer or the Army Appraisal Board.
>
> Subject to the approval of the Army Appraisal Board, the determination of just rentals for requisitioned real property, and the monthly payment of those rentals from funds of the German Government.
>
> The submission to the R. R. and C. Officer of monthly reports of activities; the reports included the new rentals determined, the rentals and billeting lists paid, and the value of the furniture and equipment supplied and of the repairs and installations made.

All requisitions of real property and its equipment held by the R. R. and C Officer were cancelled on December 1, 1920, and the property held under those requisitioned by the German Billeting Commission and transferred on temporary receipt to the American organizations or individuals concerned. Thereafter all additional property required by the A. F. G. was requisitioned by the German Billeting Commission.

The Army Appraisal Board continued in existence, but its functions were reduced to those of a board of inquiry and control. It determined the needs of the A. F. G. for real property and its equipment, maintenance and operation, and directed the German Billeting Commission to supply those needs. In addition, it investigated and adjusted the complaints of individual owners of requisitioned property.

Because of the depreciation of the German currency, it was found in many cases that the rentals previously determined were inadequate. A re-determination of such rentals was made, the increased costs of maintenance and operation being taken into consideration. The appeals of owners of requisitioned property, relative to the rentals paid them, were considered and adjusted by the Billeting Board.

<u>REQUISITIONS</u>[1]

Requisitions were classified according to their objects, as follows:

> Real property
> Furniture, furnishings, repairs, etc.,
> Services.

The requisitioning of real property[2] such as Government buildings, public utilities, hospitals and private residences, was done with due regard to the needs of the Germans, and only the property actually necessary for the A. F. G. was requisitioned. The burgermeisters of the garrisoned towns were required to furnish

1 See Instructions for the R.R.& C. Service at the end of this chapter.

2 Ibid - Section III, Pars. 9-15.

lists of all private residences, showing the number of rooms in each and the number of people in the family of the householder. These lists were carefully examined, and as far as practicable there were requisitioned in private residences only those rooms that could be spared without undue hardship to the families concerned. The billeting situation was particularly difficult in Coblenz which was over-crowded; in a few instances it was necessary to requisition entire houses, but in such cases other accomodations were provided for the German owners.

The requisitioning of furniture, furnishings, repairs, etc.,[1] was confined to the legitimate needs of the A. F. G. and the members thereof. In order to control the matter, requests from members of the A. F. G. for furniture, furnishings, etc., were submitted to the Town Majors for decision as to whether requisitions should issue. A list of articles considered reasonably necessary for the comfort and convenience of the occupants of billets was drawn up and approved by the American Headquarters; requests for furniture, etc., were checked against this list, and those found excessive were disapproved. The requisition of such luxuries as pianos was never permitted. In some instances, through ignorance or otherwise, officers made unauthorized purchases of household articles and sent the bills to the Town Majors, requesting that they be covered by requisition. This practice was discontinued upon the issue of Bulletin No. 155 on November 21, 1920, which reads as follows:

<u>REQUISITIONS ON GERMAN AUTHORITIES</u>

Authorized furniture or furnishings for quarters, billets,

[1] Ibid - Section III, Pars. 16-23.

messes, etc., may be obtained only by application to the local Town Major. Such application will be in writing and will state in what premises the articles are required. In the City of Coblenz the Town Major will receive applications only on prescribed requisition forms signed by the officer concerned. Forms may be obtained at the Town Major's office.

Articles so furnished are assigned to the premises for which they are furnished and will not, under any conditions, be removed from these premises by any other than the Town Major or his representative.

There have been several instances of officers purchasing furnishings and then forwarding the bills for same to the Town Major, with the request that the purchases be covered by a requisition. This practice is contrary to existing regulations and will be discontinued. In any instances of this nature the officer responsible for the purchase will be required to pay for the same from his own funds.

Prior to the reduction of the A. F. G. requisitions for Services[1] were issued on very few occasions, and then only because the actual necessity had arisen and the requisitioning had been approved by Headquarters.

Requisitions of every kind were formulated in the offices of the various Town Majors and forwarded through the Zone Majors to the central R. R. and C. office for approval. They were issued in quintuplicate, copies being distributed to the central R. R. and C. office in Coblenz, the German Billeting Commission and the Town Majors in whose offices the requisitions originated. An accurate record of the costs of fulfilled requisitions was kept in the central R. R. and C. office, which rendered a monthly report thereof to the Auditor, A. F. G.

As in the case of real property, it had been customary to secure furniture, furnishings, etc., by requisition upon local burgermeisters. This practice also was discontinued on December 1, 1920, and thereafter all such articles were secured through the German Billeting Commission.

[1] See Instructions for R.R. and C. Service - Section III, Pars. 4-8

CLAIMS[1]

A letter from the Commander-in-Chief of the A. E. F., under date of May 1, 1919, prescribed a provisional policy for the settling of claims for damages caused by American troops in the American Area. The policy was based on that followed in France, sufficient changes being made to meet the special conditions existing in an occupied territory.[2] With this letter as a guide, a new policy was then adopted by the Third Army[3]; this policy, as amended from time to time to cover practical details[4], continued in force until November 1920 when it was modified as will be shown later on.

It often happened that claims submitted by the Germans were disallowed. In January 1920 it was decided to issue "disallowal notices" to cover all claims which, either wholly or in part, were not considered to be proper charges against the United States.[5] The disallowal notice contained a brief description of the claim, the amount of the claim, the reasons for the disallowal, and other pertinent information. The notice was sent to the claimant who was privileged to submit a new claim to the German Government, provided the disallowal notice was attached. After verification, through the R. R. and C. office, by the German authorities, the voluntary payment of the renewed claim by the German Government was authorized, but with the distinct understanding that thereby the United States had been relieved of

1 See A. M. G., Volume III, Page 735, and Section IV, Instructions for R.R. and C. service at the end of this chapter.
2 A. M. G., Volume III, Page 744.
3 Ibid, Page 747.
4 Ibid, Page 750.
5 Ibid, Page 760.

all responsibility in the matter. In other words, the amount so paid would not be set off against the costs of occupation, which were a charge against the German Government.

Before the adoption of the policy outlined above, thousands of claims had been submitted to the German Government; these claims were sent by the Oberprasident of the Rhine Province to the R. R. and C. Service for investigation. Such as were found to be duplicates of claims already disallowed by the R. R. and C. Service were returned to the German authorities with authority for their payment by the German Government. Those claims that had been approved in part by the R. R. and C. Service were returned to the German authorities with permission to pay amounts equal to the differences between the amounts originally claimed and the amounts allowed by the United States. Claims that never before had been submitted to the American authorities were forwarded to the R. R. and C. representatives in the indicated Kreise for investigation and recommendation.

During the summer of 1920 the American troops stationed in the outlying districts of the Coblenz Bridgehead were gradually withdrawn and concentrated in Coblenz and vicinity and in the Kreise of Mayen and Neuwied. By this time the claims from the outlying districts had been settled or otherwise disposed of, and in the autumn of 1920 the offices of the Zone Majors in the Kreise of Cochem, Adenau, Ahrweiler, Neuwied and Montabaur were closed.

The methods of dealing with claims in accordance with the policy prescribed in the letter of May 1, 1919, from the General Headquarters of the A. E. F., was modified in November 1920. It became necessary to discontinue the existing methods of handling

Class "A" claims[1], as on August 10, 1920, the Judge Advocate General of the Army had issued the following opinion with regard to claims arising in Germany:

> As I am convinced, for reasons that will presently appear, that these provisions for the payments of damages to inhabitants of Germany are altogether illegal, I have thought it interesting to inquire how they came to be written into the regulations.
>
> Whether, under the laws of Germany, that Government was or is responsible to compensate its nationals for losses and damages inflicted by an invading army, is a matter upon which I do not feel competent to speak; but so far as the Interallied instructions include such losses and damages as are not inseparably bound up in the right of requisition within the obligations of Article IX of the Armistice, I think they proceeded upon a mistaken theory; and so far as those instructions impute to the Government of an occupying army the duty to compel payment for such losses and damages from the German Government to its own nationals, I am convinced that the Interallied position was fundamentally unsound from the viewpoint of international law.
>
> There is no authority of law, municipal or international, for any such proceedings as those prescribed in the R.R.& C. Instructions. Except within the narrow limits of the 105th Article of War, there is no authority of law for the payment out of public funds of the U.S. for any damages, as such, to the inhabitants of enemy territory, whether occupied or unoccupied. There is no authority of law, written or unwritten, conventional or customary, for compelling payment by the enemy state to its own nationals on account of damages, purely as such, suffered at the hands of the occupying army. No such right exists under laws and customs of war, nor is any such duty to protect an enemy inhabitant of occupied territory against his own national government cast upon an invading army. If any such right or duty existed in the present situation, it must have been created by agreement with the German Government, that is to say, by some stipulation of the Armistice Convention.
>
> By the second sentence of Article IX (of the Armistice Convention), the German Government assumed a certain definite burden, namely, the expense of maintaining the Army of Occupation. The word in the original Armistice document is 'entretien' - It means maintenance, that is to say support, upkeep. It includes subsistence, the lodging, the clothing and, as I think, the pay of the troops. There may be some difference of opinion as to its precise limitaion, but there can be no possible doubt that it does not include compensation for losses and damages inflicted upon the inhabitants by the occupying army. The compulsion of military authority which forces the payment from German public funds for losses and damages inflicted by the Armerican Army is not in violation of international law but —is a maladroit proceeding. The claims— are investigated by American authority; the fault of the American Forces conceded —; the liability of the U.S. determined —; the claims vouchered — as a just indebtedness of the

[1] A.M.G. Page 744, Volume III.

U. S.; the evidence of all this placed in the hands of the private enemy claimant and the public official; the military order to make payment out of German public funds and finally the witnessing and certification of American officials that such payment has in fact been made.

Beyond these payments (requisition or negotiated purchase), whether out of German funds in the proper case or out of U.S. funds where the German obligation falls short of covering them, no payments can lawfully be made to enemy inhabitants of any part of Germany, for losses or damages inflicted by the American Army, except always such as may be compensated by operation of the 105th Article of War. For the rest, all claims for damages and losses beyond the limits just defined, should be received and promptly and thoroughly investigated; the facts should be ascertained and the evidence assembled and preserved, but the process should end there. There should be no determination of liability on the part of the U.S. or its officers or soldiers, no liquidation of damages, no vouchering of the claim as an obligation of the U.S. and no delivery to the claimant or any other enemy person of the results of the inquest, much less of a certificate of indebtedness. Above all it should be conceived that whatever is done, is done solely for the protection of the U.S. (for there is no other interest) and not, as the R.R.& C. Instructions have it, "in order that the individual (enemy) citizen may be protected".

This opinion was transmitted to the A. F. G., and after due consideration the Commanding General addressed The Adjutant General of the Army, in part, as follows:

The German authorities are required by these Headquarters to pay Class "A" claims out of German funds. No funds of the United States are employed for any such purposes, so that it follows that the objections raised by the Judge Advocate General must be limited to the fact that these Headquarters have been requiring the German authorities to pay such claims, and to the method by which this has been done.

To require the American Government to pay such claims would require it to pay a part of the cost and upkeep of the army of occupation, contrary to the express stipulations of the Armistice, but it has always been believed by these Headquarters that some kind of moral and equitable duty rests upon the Commanding General of the American Forces in Germany to make it possible for the German Government to do justice to its citizens in connection with the damages growing out of the occupation by our troops. It is altogether impossible to permit the German authorities to make an investigation of such claims by summoning our soldiers or inspecting the premises used by our forces for military purposes. The result is that unless the same are inspected by the American authorities and the amount of damages communicated to the Germans the latter could never act intelligently on such claims. Furthermore, nothing is admitted except the truth as established by our own investigating officers, that is, that the damage was caused by our forces as incident to our occupation.

In view, however, of the opinion of the Judge Advocate General, certain modifications of this procedure seem necessary, and the following recommendations are submitted:

(a) That Class "B" claims be continued to be investigated, reported upon and filed together with all of the relevant testimony obtained thereon.

(b) That the 105th A.W. be applied as heretofore to Class "C" claims.

(c) In regard to Class "A" claims, that the same be investigated as heretofore and that the result be communicated to the German authorities, but that instead of compelling them to pay the same they be notified that the R.R.& C. Service has investigated the claim and found that it resulted as an incident to our occupation and as the necessary consequences of the functioning of our troops, but through no negligence or fault of any individual belonging to the American Forces; that in the opinion of the R.R.& C. Service the claimant is entitled to a certain amount of damages which should be paid by the German Government, but that all liabiltiy to the same is expressly disclaimed by the American Government.

(d) That the form of voucher now being used, **citing the United States as a debtor to the German national, be discontinued; that the practice of authorizing the German Government to pay claims disallowed by these Headquarters be also discontinued.**

That in cases where all the other facts are the same, no distinction be made between injuries to the person or to the property under Class "A" claims.

The War Department approved these recommendations on March 23, 1921. **Thereafter** the notices to claimants covered the amounts that had been approved as being incident to the occupation. A copy of the notice was sent to the German authorities, and the decision as to payment was left to the discretion of the German Government. The notice was worded as follows:

Relative to your claim No._______ submitted on (date), you are hereby advised that same has been investigated, and it has been found that the damage in question is the result of, and incident to, the occupation of the American Forces, and as the necessary consequences of the functioning of our troops, but through no fault or negligence of any individual belonging thereto.

In the opinion of these Headquarters you have suffered damages amounting to_______ Marks. It is recommended that you request payment of your claim by the German Government with the understanding that all liability for same is expressly disclaimed by the United States Government.

The practice of sending a disallowal notice to one whose claim had been disapproved was discontinued also. The Judge Advocate of the A. F. G.

then submitted the following opinion:

> The writer's attention has been invited to Par. 5, Section 4, Instructions for the R.R.& C. Service, which provides that "No disallowal notices will be made out by the Zone or Town Majors". It is believed that this provision is a result of an opinion of the Judge Advocate General with reference to the handling of claims put forward by German civilians, in answer to which it was recommended that the practice of giving written permission to the German Government to pay disallowed claims be discontinued. There is nothing in the opinion of the Judge Advocate General, and in the opinion of this office there is nothing in principle, against notifying German claimants when their claims are disallowed. On the contrary, it is believed to be a a matter of courtesy and fairness to give them a written notice, stating that the claim has been investigated by the American authorities and disallowed.

This opinion was approved by the Commanding General, and thereafter the following notice was sent to those whose claims had been disapproved:

> Your claim for damages for________Marks has been referred to these Headquarters, and upon investigation it is our opinion that no damage has been sustained.

CLAIMS UNDER ACT OF APRIL 18, 1918

No provision had been made by the War Department for the investigation and payment, subsequent to the dissolution of the A. E. F., of claims arising under the Act of Congress of April 18, 1918. On January 10, 1920, the Commanding General of the A. F. G., by cablegram to the Secretary of War, requested that he be designated by the President to receive and approve claims for payment under that Act and, in the name of the Secretary of War, to issue such regulations governing their payment as might be deemed necessary. On February 10, 1920, the request was approved by the Secretary of War. The R. R. and C. Service was then reorganized so that its scope might include all claims arising under the Act mentioned, with the exception of those in which the responsibility had been transferred by contracts made by the United States Liquidation Commission. The reorganization was made effective by General Orders No. 21, issued on

March 3, 1920, which read as follows:

1. An Act of Congress, approved 18 April 1918, provides as follows:

AN ACT TO GIVE INDEMNITY FOR DAMAGES CAUSED BY AMERICAN FORCES ABROAD.

BE IT ENACTED BY THE SENATE AND HOUSE OF REPRESENTATIVES OF THE UNITED STATES OF AMERICA IN CONGRESS ASSEMBLED. That claims of inhabitants of France or of any other European country not an enemy or an ally of an enemy for damages caused by American military forces may be presented to any officer designated by the President, and when approved by such an officer shall be paid under regulations made by the Secretary of War.

Section 2. That claims under this statute shall not be approved unless they would be payable according to the law of practice governing the military forces of the country in which they occur.

Section 3. That hereafter appropriations for the Incidental Expenses of the Quartermaster Corps shall be available for paying the claims herein described.

Section 4. That this statute does not supersede other modes of indemnity now in existence and does not diminish responsibility of any member of the military forces to the person injured or to the United States.

2. Pursuant to authority received by War Department cable No. 134-W, paragraph 1, signed HARRIS, and dated 10 February, 1920, the activities of the Renting, Requisitions and Claims Service, Headquarters, A.F.G., are extended so as to include the receiving and payment of claims under the foregoing Act, and the making and publishing in the name of the Secretary of War subject to the approval of these Headquarters - regulations to carry out the provisions of this Act.

3. The administration of the Service shall devolve upon the following officers:

- (a) A Chief Officer whose title shall be Director of the Renting, Requisitions and Claims Service.
- (b) An R.R.& C. Officer for the German Occupied Area.
- (c) An R.R.& C. Officer for the Port of Antwerp.

4. The principal duties of the Renting, Requisitions and Claims Service shall be as follows:

- (a) For the R.R.& C. Officer in German Occupied Area as prescribed in the Instructions for the R.R.& C. Service, Headquarters, A.F.G., dated 17 September 1919, which are continued in force.
- (b) For the R.R.& C. Officer, Port of Antwerp, as far as conditions permit, as prescribed in the Manual for R.R.& C. Officers, approved 31 August 1918, modified by War Department cable No. 134-W, paragraph 1, dated 10 February 1920, signed HARRIS.

5. The Renting, Requisitions and Claims Service is authorized to correspond directly with all officers in matters within its jurisdiction. Its requests for information will be complied with as promptly as possible. Correspondence between the R.R.& C. Officer, Port of Antwerp, and the Director of the R.R.& C. Service will be direct.

6. The findings and directions of the R.R.& C. Service, and all its authorized representatives, properly authenticated, will constitute authority for payment of all rents, requisitions, billets, allowances and claims - subject to review by the Commanding General, A.F.G., only in cases cited below.

7. The R.R.& C. Officer, Port of Antwerp, has jurisdiction in all matters of renting, requisitions, billeting and all claims payable under the Act of 18 April 1918. He has authority to approve claims for payment in amounts not to exceed 2,500.00 francs. In the case of Admiralty claims he will investigate them, make his recommendation in each case, and forward them to the U.S. through the office of the Director, R.R.& C. Service, for settlement. Claims of inhabitants of France for damages occurring before 1 December 1919, will be referred to the French Government for settlement.

8. Claims in excess of 2,500.00 francs and under 10,000.00 francs will be approved by the Director, R.R.& C. Service. Claims in excess of 10,000.00 francs will be approved by the Commanding General, A.F.G.

9. The Disbursing Quartermaster, Port of Antwerp, is authorized to pay all vouchers approved by the R.R.& C. Officer, Port of Antwerp, for matters within the latter's jurisdiction.

10. Neither the responsibility of the wrong-doer, or the duties of the Commanding Officer, under the 105th Article of War, is affected by the enactment of the law promulgated in this Order, and no payment of claims will be made from public funds where the provisions of the 105th Article of War can be so applied as to compensate the claims. To prevent duplication of claims a stoppage within the 105th Article of War shall be immediately reported by the officer ordering stoppage to the Director, R.R.& C. Service, Headquarters, A.F.G., which report shall include a statement of the name and address of the person in whose favor the stoppage is ordered, the amount of the stoppage and a brief description of the claims upon which it is based.

11. Major Jno. B. Hogan, C.A.C., is hereby detailed as Director, Renting, Requisitions and Claims Service, (R.R.& C.) in addition to his other duties as R.R.& C. Officer, German Occupied Area. Major J. K. White, J.A.G. Department, is designated as R.R.& C. Officer, Port of Antwerp, in addition to his other duties. The Director, R.R.& C. Service shall - subject to the approval of the Commanding General, A.F.G. -make such rules and regulations and issue such instructions to the officers of the R.R.& C. Service as may be necessary.

The Director, R.R.& C. Service, was also the R.R.& C. Officer for the occupied portion of Germany. In his capacity as Director he was authorized to receive and settle all claims arising under the Act of April 18, 1918, except those in France, Italy and Monaco. The R.R.& C. Officer for the Port of Antwerp was authorized to receive and pay all claims arising in Belgium.

Much difficulty had been experienced in the settlement of the claims of Europeans against the United States; this was due largely to the de-centralization of the recording of such claims. Several claims had been sent from Washington to Coblenz, with instructions that they be paid; investigation showed that in two or three of these cases the claims had already been paid, although no record of payment could be found in the R.R. and C. Office at Coblenz. To avoid duplicate payments, arrangements were made to have the records centralized in the Coblenz office, thereby eliminating the danger of duplication. These arrangements, which had been made in July 1920 in conference between the Director of the R.R.& C. Service of the A.F.G. and the R.R.& C. Officer at Paris, and had been approved by the Commanding General of the A.F.G., were as follows:

> All claims, of whatever nature, arising in France, Italy or Monaco, and which, therefore, came under the contracts made by the United States Liquidation Commission, were to be settled through the R.R.& C. Officer in Paris. All Graves Registration Service claims arising in any country other than Germany and all Admiralty claims referred to the Paris office from Washington, or presented to the Paris office for reference to Washington, were to be handled through that office.
>
> All claims arising in Germany were to be handled through the R.R.& C. Officer of the A.F.G. All Graves Registration Service claims arising in Germany were to be sent to the Headquarters of the Graves Registration Service in Paris for investigation and recommendations; the claims would then be returned to Coblenz for approval (or disapproval) and settlement in accordance with existing regulations. All claims, except those pertaining to the Graves Registration Service, arising in any European country other than France, Italy and Monaco, were to be forwarded to the R.R.& C. Officer of the A.F.G. for final settlement and disposition.

LUXEMBOURG AGREEMENT

On April 21, 1921, there was completed a contract whereby the Grand Duchy of Luxembourg, for a specified sum, agreed to settle all claims of whatever nature that remained outstanding against the United States. The arrangements were made in conference between G-1 and the Direcotr of the R.R.& C. Service on one side and representatives of Luxembourg on the other. The contract was signed for the United States by the Judge Advocate of the A.F.G., who acted for the Commanding General, and by the Minister-President of the Grand Duchy for Luxembourg.[1]

The contract was, in part, as follows:

> Luxembourg hereby substitutes itself for the United States in reference to and guarantees the latter against all claims, demands, obligations, rights, actions, suits and controversies, of every kind and nature, arising in Luxembourg during the period stipulated in this agreement, which it, the said Duchy of Luxembourg, or any inhabitant, or any person subject to the sovereignty of Luxembourg, or any organization, association or corporation, commune or department of Luxembourg, may now have or which may hereafter accrue for damage to, use or loss of property, or injuries to or death of persons, caused by or resulting from acts, or omissions, of members of the said American Military Forces, or of any one connected therewith, during the period embraced in this agreement, whether the same are listed below or not, but with particular reference to the following:
>
> (a) All claims which the United States is authorized to pay to the inhabitants of Luxembourg, known more explicitly as "private claims".
>
> (b) All claims for the use and occupation of railroads, cars, tracks, and for transportation of troops, supplies, and material.
>
> (c) All claims commonly called "Commune Claims" for damages to boundary roads, highways, waterways, etc.
>
> (d) All claims of the Department of Highways, Public Works, and Finance, Luxembourg, for damages to highways and public works.
>
> (e) All claims for billets and offices and like accomodations, in accordance with the agreement as to payment hereinbefore mentioned.
>
> The specifying of the particular claims in sub-paragraphs (a) to (e), both inclusive, shall not be construed to exclude any

1 The three American officers were decorated by the Luxembourg Government.

other claims described in Article I above, but to the contrary, the intention of this agreement is to protect and hold harmless the United States against all claims of every kind and description, whether the same be contractual or tortuous, arising or growing out of any acts of negligence on the part of any of its said Forces in Luxembourg during said period.

The United States agrees to pay Luxembourg 2,518,381.17 Luxembourg francs, or the equivalent in French francs, upon the signing of this agreement.

OFFICERS

The R. R. and C. Service operated under the supervision and control of G-1. The following officers served, respectively, as R. R. and C. Officer, A. F. G., during 1920 and 1921:

Major John B. Wogan, C.A.C.,
Major C.M.S. Skene, C.A.C.,
Captain Charles C. Bodeker, Inf.,
Captain Stephen Mabon, F.A. (Assistant)

The following officers were among those who served as Zone and Town Majors or as Investigating Officers for the Army Appraisal Board:

Major C. N. Feamster, Inf.,
Major William Fiske, Inf.,
Captain F. M. Distlehorst, Inf.,
Captain W. H. Donaldson, C.A.C.,
Captain H. C. Fellows, Cav.,
Captain T. F. Fieker, Inf.,
Captain H. M. Herff, Inf.,
Captain M. A. S. Ming, F.A.,
Captain H. W. Schwartz, Inf.,
Captain L. J. Vogel, F.A.,
Captain O. K. Wolber, Inf.,
1st Lieut. L. M.Hanna, C.A.C.

MISCELLANEOUS

The change in the policy governing requisitions and rentals, which was made in December 1920 at the request of the German Government, is shown in the following General Orders:

HEADQUARTERS
AMERICAN FORCES IN GERMANY
COBLENZ, GERMANY

2 December 1920.

GENERAL ORDERS
NO. 104

BILLETS AND REQUISITIONS OF SUPPLIES AND PROPERTY,
UNDER R. R. AND C. SERVICE.

On and after 1 December, 1920, the payment of rentals of all billets and requisitioned real property will be made by the Central German Government, acting through the German Billeting Commission, whose appointment has been approved by these Headquarters.

After 1 December, 1920, no further requisitions for the use of real property will be made by the R.R. & C. Service. All property formerly obtained on requisition by this Service will now be obtained by the German Billeting Commission. The use of War Department Voucher 330 and Field Form No. 1 will be discontinued.

When it is desired to have real property requisitioned, a letter in quadruplicate will be forwarded to the R. R. & C. Officer, these Headquarters, by the Town Major concerned. The following is an example of the form letter to be used:

"File No. ____________ Place ____________________
(To be entered at Hqrs.R.R.& C.) Date ____________________

It is desired that the German Government requisition without delay for the exclusive use as quarters for a commissioned (non-commissioned or I.A.R.R.C.) officer and family of the American Forces:

Three rooms, joint use of kitchen and bath, in the House at No. 40 Hochstrasse, Coblenz, the property of Herr Rolf A. Schmidt, including maintenance and upkeep, the necessary equipment and the necessary amount of heat, light and water.

Occupancy to begin on (date) __________

Approved: ____________________
Town Major

____
R.R.& C. Officer, A.F.G.

1st Ind.

R.R.& C. Office, Headquarters, American Forces in Germany, 2 December 1920. To: The President of the German Billeting Commission....."

The quadruplicate copy of this letter will be returned to the Town Major after it has been approved and the R. R. & C. file number entered.

Conforming to the above change, parts I, II and III of Instructions for the R. R. & C. Service, dated 17 September, 1919, are revoked and the following substituted therefor:

I. GENERAL.

1. Local troop commanders will conduct all their business with the German authorities for requisitions, billets, etc., through the local representative of the R. R. & C. Service. The latter will avoid as far as possible dealing directly with the local inhabitants. This, however, will not apply in the case of claim investigations when local inhabitants have to be questioned in order that evidence may be obtained.

2. The local R. R. & C. representatives are charged with the following:

(a) Making arrangements for billets.
(b) Making arrangements for the requisitioning by the Germans of all real property, supplies and services, when request has been made for same and approved by proper authority.
(c) Facilitating the settlement of all claims or demands of the Germans arising from the occupation of their area by American Troops.

3. A Zone Major has been appointed for each brigade in the American occupied territory. A Town Major, or Acting Town Major, will be appointed in each town, village or Gemeinde in which troops are billeted.

4. All correspondence between the R. R. & C. Office, Headquarters, American Forces in Germany, and the Town or Acting Town Major will be through the Office of the Zone Major - except in the case of the Town Major of Coblenz.

5. Instructions contained in this memorandum for Zone Majors and Town Majors apply equally to Acting Zone Majors and Acting Town Majors.

II. BILLETING.

1. The Town Major will call upon the Burgermeister for a list showing the town's billeting facilities, and he will keep a complete and accurate list of all such accomodations, showing for each building its capacity for officers, men, offices, storerooms, animals and other needs to be anticipated.

2. The Town Major will keep a record of billets and the organizations which occupy them at different periods, to be used in adjusting claims submitted after the units have left the area.

3. The commanding officer of the troops is responsible for the location, maintenance and operation of suitable latrines, baths, water, washing and incinerating facilities. He will arrange for the disposal of refuse, for fire protection and for traffic regulations whenever necessary. He will consult with the Town Major in making arrangements in this and similar matters. He will establish directory signs at all points of the village where they will be advantageous.

4. Billets will not be changed without the authority of the commanding officer. Troops will not occupy billets not assigned to them. Changes in the personnel of an organization which effect billeting will be reported at once to the Town Major. This provision is absolutely essential in order that proper record may be kept of the amount due each householder for billeting services rendered.

5. As the damage or destruction of private property by troops is usually the basis of subsequent claims, troops will be cautioned in this respect. There will be no molestation of inhabitants, and no entering of private premises, except those assigned as billets. When property is in a damaged condition prior to its use by the troops, this will be noted by the Town Major to prevent the subsequent allowance of improper claims.

6. The Town Major upon request, and in the manner described below, will provide dining rooms and kitchens for officers' messes. Services and supplies paid for by the troops will be covered by private agreement.

7. Troops billeted or quartered among the inhabitants or in localities used as domiciles have a right to lighting, heating and a supply of water in common with the inhabitants.

Method of Payment.

8. At the end of each calendar month, or at the end of a period of occupancy in the case of a shorter stay, each Town Major will prepare a billeting distribution list showing the total number of billets for officers, enlisted men, animals, etc., furnished by the Town and for what period they were furnished.

9. The billeting distribution lists will be prepared in triplicate. Each copy will be signed by the Town Major and the Burgermeister, and stamped by the latter with the seal of the town. The original and the duplicate will be

forwarded to the R. R. & C. Officer, these Headquarters, and the triplicate will be retained by the Town Major for his file. The duplicate copy will be forwarded to the German Billeting Commission which will settle upon the payment to be made with the owners of the property occupied.

10. Separate B.D.L.'s will be prepared for each town, village or Gemeinde in which troops are billeted.

11. The billeting certificate will no longer be used.

12. The name of the householder, location of the house, together with the number of rooms and the period during which they were occupied as messes, kitchens, orderly rooms, company, detachment, supply and guard rooms will be shown on the B.D.L.'s. Any supplies or services received, which it has been the practice for Town Majors to requisition in the past (not paid for by the troops) will also be shown on the B.D.L.'s.

III. REQUISITIONS.

Requisitions for Supplies and Material.

1. The procurement of supplies and material by requisition locally will be governed by the following:

> Local commanders are authorized, without prior approval of higher authority, to enforce requisitions on local German authorities for fuel alone, and then only to meet requirements for authorized issues within regulation allowance. This authority they do not have if such supply is maintained from established sources of supply or from collective requisitions which may be demanded by these Headquarters from German authorities. This authority to requisition without the approval of higher authority is limited to fuel and only under the conditions imposed in this paragraph.

2. The procurement of all other classes of authorized supplies required locally will follow the rules established in this memorandum. No food or forage will be requisitioned without the prior approval of Headquarters, American Forces in Germany, and then only in quantities sufficient to meet an emergency.

3. Quantities of fuel and forage will be expressed in kilos on all requisitions. For all other classes of supplies the ordinary U. S. classification and description will be employed.

Requisition of Services.

4. Local commanders are not authorized to enforce

requisitions for services without the approval of these Headquarters.

5. It is contrary to the announced policy of our Government to make direct requisition for labor except in emergency. No direct requisition, therefore, will be made for labor without the prior approval of these Headquarters. In the event of such approval the method of payment will be described.

6. When public buildings and utilities, such as Government-owned buildings, barracks, depots, hospitals, water-works, electric plants, etc., which already have a working force such as janitors, charwomen, engineers, mechanics, etc., are taken over for the use by the American Forces under a German requisition, the requisition will include the ordinary and necessary personnel for their operation. The proper German official will be required to pay these employees out of public German funds at the same rate as they were paid prior to our occupation. It is incumbent upon all U. S. officers in charge to see that these payments are promptly made.

7. When private buildings and utilities, such as private owned hotels, private hospitals, factories, etc., which already have a working force, are taken over by a German requisition, and when it is necessary and has been authorized to retain such force, the following procedure will be observed:

> Individuals included in this working force will be paid at the same rate as previous to our occupation, either from U. S. funds or by the German Government. The letter asking that the place be requisitioned will be so worded as to show this agreement.

8. In all cases where supplies or services, as distinguished from real property, are requisitioned, requisition blank Form 1 will be properly executed in duplicate and forwarded direct to Headquarters, American Forces in Germany. These requisitions must be so drawn as to permit the verbage on the face of the requisition being used as the complete voucher upon which subsequent settlement may be made.

Requisition for the use of real property.

9. The German Government, through the German Billeting Commission, will requisition the use of all real property, formerly requisitioned by this service, in the American occupied area after 1 December, 1920. In order to have real property requisitioned the following steps will be observed:

(a) The service or party interested, in conjunction with an R. R. & C. representative, will locate the property.

(b) The R. R. & C. representative will forward a

letter in quadruplicate to the Director, R. R. & C. Service, American Forces in Germany, requesting that the property be requisitioned.

(c) The R. R. & C. Officer, will have the letter requesting a requisition approved and the file number entered. After approval the original will be filed in the R. R. & C. Office; the duplicate and triplicate will be forwarded to the German Billeting Commission by indorsement, and the quadruplicate will be returned to the R. R. & C. representative who originated the letter. Upon receipt of this copy the R. R. & C. representative will notify the interested service that the request for requisition has been approved and that the premises may be occupied; he will also notify this office of the date of occupancy.

(d) The rental to be paid will be determined by the German Billeting Commission with the approval of the A. A. B., American Forces in Germany. Payment of rentals will also be arranged by the German Billeting Commission.

10. Requests for requisitions for real property to be used by the I.A.R.H.C. or other Inter-Allied Commissions will be forwarded in quintuplicate. The first four copies will be disposed of in the same manner as outlined in sub-paragraph (b) paragraph 9 above, and the fifth copy will be forwarded by the R. R. & C. Officer, to the organization interested.

11. In order to reduce the claims, which no doubt will be presented upon the evacuation of quarters requisitioned for the use of American officers, the following procedure will be followed:

> When the request for the requisitioning by the Germans of any quarters for the permanent occupancy of American officers and others is approved, the owner of the quarters will be required to submit to the Billeting Officer, in triplicate, a list of all movable furniture and equipment turned over with the quarters. The officer or person who occupies the quarters will be required to check this list, certifying to its correctness upon the face. A copy of the list with the officer's certification thereon will be filed with the letter requesting the requisition of the billet in the R. R. & C. Office. The officer occupying the quarters will retain one list in his possession until he vacates the quarters, at which time he will make a final check of the articles in the quarters and turn this list in to the Billeting Office, who will forward said list to the R. R. & C. Office, these Headquarters, noting thereon the condition and whether or not any of the articles are missing.

Cancellation of requisitions on real property.

12. Steps will be taken to cancel promptly the German requisitions on real property which has been requisitioned at

our request, when the use of such property is no longer needed. This will be done as follows:

(a) The service or party evacuating the premises will notify the local R. R. & C. representative to that effect. The local R. R. & C. representative in making this report will give the file number on the returned copy of the letter which requested that the property be requisitioned in the first place and the date that the requisition should be cancelled. The German Billeting Commission will be notified by these Headquarters that the use of the property is no longer desired and that the requisition may be cancelled.

(b) In cases where the evacuating service or party fails to notify the local R. R. & C. representative, the latter will, as soon as he becomes cognizant of the fact that the premises have been evacuated, find out the date upon which they have been so evacuated and proceed as in sub-paragraph (a) above.

13. Orders, bulletins and instructions heretofore issued, which conflict with the provisions of this order are hereby revoked.

(G-1. R.R.& C.)

By command of Major General ALLEN:

OFFICIAL:
S. G. TALBOTT,
Adjutant General.

WILLIAM H. JOHNSTON,
Chief of Staff.

The following Instructions for the R. R. & C. Service were published on 1 May 1921 by Headquarters, A. F. G., for the information and guidance of all concerned:

INSTRUCTIONS FOR THE R. R. & C. SERVICE

I. GENERAL

1. The following instructions are intended to cover in detail all activities of the R. R. & C. Service. They will be carefully studied and rigidly complied with by all concerned. In compiling these instructions, the R. R. & C. Officer has endeavored to omit any surplus instructions or reports so as to avoid, as much as possible, the red tape usually connected with a service of this sort.

2. Local troop commanders will conduct all of their business

with the German authorities for quarters, billets, other real property and the equipment thereof, through the local representative of the R. R. & C. Service. The latter will deal with the local burgermeister or the local representative of the German Billeting Commission, as the case may be, and will avoid as far as possible dealing directly with the local inhabitants. This, however, will not apply in the case of claims investigation, when local inhabitants must be questioned in order that evidence may be obtained.

3. Local R. R. & C. representatives are charged with the following:

(a) Making arrangements for billets.
(b) Requesting the requisitioning of all real property, supplies and services, when request for same has been made and approved by proper authority.
(c) Facilitating the settlement of all claims or demands of the Germans arising from the occupation of their area by American troops.

4. A Zone Major (Z. M.) will be appointed by these Headquarters for each Brigade Area (or Division Area, in the event of Division organization), in the American Occupied Territory. A Town Major (T. M.) or Acting Town Major (A. T. M.) will be appointed in each town, village or Gemeinde in which troops are billeted.

5. All correspondence between the R. R. & C. Office, Hq. A. F. G., and the T. M. or A. T. M., will be through the office of the Z. M., except in the case of the T. M. of Coblenz.

6. Instructions contained in this memorandum for Z. M.'s and T. M.'s apply equally to A. Z. M.'s and A. T. M.'s.

II. BILLETING

General.

1. The T. M. will call upon the burgermeister for a list showing the town's billeting facilities, and he will keep a complete and accurate list of all such accommodations, showing for each building its capacity for officers, men, offices, storerooms, animals, and other needs to be anticipated. He will also keep a record of the water supply, giving amount, location and quality.

2. The T. M. will keep a record of billets and the organizations which occupy them at different periods, to be used in adjusting claims submitted after the units have left the area.

3. The commanding officer of troops is responsible for the location, maintenance and operation of suitable latrines,

baths, water and of washing and incinerating facilities. He will arrange for the disposal of refuse and for fire protection whenever necessary. He will consult with the T. M. in making arrangements in this and similar matters. He will establish directory signs at all points of the village where they will be advantageous. In the event of temporary occupation of billets by members of his command at other than his permanent station, he will be responsible for the preparation of the necessary B.D.L.'s and the forwarding of the same to the Zone Major of the Area in which billets were occupied.

4. Billets will not be changed without the authority of the commanding officer. Troops will not occupy billets not assigned to them. Changes in the personnel of an organization which effect billeting will be reported at once to the T. M. This provision is absolutely essential in order that proper record may be kept of the amount due each householder for billeting service rendered.

5. As the destruction or damage of private property by troops is usually the basis for subsequent claims, troops will be cautioned in this regard. There will be no molestation of inhabitants, and no entering of private premises except those assigned as billets. When property is in a damaged condition prior to its use by troops, this will be noted by the T. M. to prevent the subsequent allowance of improper claims.

6. The T. M., upon request, and in the manner described below, will provide dining rooms and kitchens for officers' messes. Services and supplies, paid for by the troops, will be covered by private agreement.

7. Troops billeted or quartered among the inhabitants or in localities used as domiciles have a right to lighting, heating and a supply of water in common with the inhabitants.

8. COMMISSIONED AND CIVIL PERSONNEL: Military officers and civilians attached to the Army of Occupation, and their families, may be billeted on the inhabitants in accordance with these regulations, provided public buildings are not available. German civilian employees of the Army will not be billeted at the expense of the Government.

9. ENLISTED PERSONNEL: Private soldiers and N.C.O's below Grade 3, shall be accomodated in barracks, and shall not be billeted on the inhabitants, except in case of unusual circumstances, or as herein excepted.

10. Barrack space will be supplied from the following sources in the order named, conditions permitting:

a) Military compounds.
b) Buildings which are property of Central German Government.
c) Buildings which are public German property.

d) Buildings which are privately owned.

11. N. C. O.'s of Grades 1, 2 and 3 may be billeted on the inhabitants with the consent of their commanding officers, if such billets are available, and if publicly owned quarters are not available.

12. Married enlisted men and N. C. O.'s below Grade 3, may, with the consent of their commanding officers, rent quarters in private houses. In such cases the commanding officers will be held responsible for the soldier's conduct and for the payment of rent. Soldier's and N. C. O.'s of this class will not rent billets without first having consulted the T. M., who will assist them in securing quarters.

13. Where it is necessary to billet families in houses with German occupants, the T. M. will select the rooms which are to be requisitioned, and, in so far as is possible under existing circumstances, arrange that the two families occupy entirely separate quarters. When it is impracticable to give billeted families separate kitchens and baths, they will use the kitchens and baths in common with the German occupants. Detailed arrangements concerning the use of the kitchens and baths will be made directly with the German householder or through the T. M.

14. When a building is requisitioned and the German family required to move out of the building, they will be authorized to take with them only such furniture as may be approved by the T. M. Each piece of furniture will be listed and described so that there will be no confusion as to what they are authorized to move. Buildings which are not furnished, or which are partly furnished, will be completely furnished by the burgermeister. The T. M. will decide what furnishings are necessary.

How obtained.

15. Each T. M. will call upon the local burgermeister for a list of the billeting facilities of his town, and will keep a complete and accurate record of all such accomodations, showing for each building its capacity for officers, men, offices, storerooms, animals and other needs to be anticipated. The T. M. may reject any of the locations submitted to him by the burgermeister which, in the opinion of the T. M., are not acceptable for use. If the above mentioned method of obtaining billeting accomodations proves unsatisfactory, the T. M. may, through his own inspectors, search for and upon the approval of these Headquarters, requisition additional locations. All billeting accomodations accepted by the T. M. will be inspected and classified by him for assignment when needed.

16. Single billets may be obtained upon application to the local T. M., with the proper credentials. A Billeting Order will be issued by the T. M., and the billet applicant will serve this Billeting Order upon the householder whose address is given thereon. For single billets two kinds of Billeting Orders

are issued: Billeting Order "B", which entitles the applicant to quarters at Government expense; and Billeting Order "A", which entitles the applicant to quarters which will be paid for by himself. (See models in Appendix).

17. Persons whose credentials show them to be on temporary duty, will be issued a Billeting Order for the period during which they will remain in the community. Those persons who are on permanent or indefinite duty in the community, will be given a permanent Billeting Order. In either case, upon vacating a billet, the occupant will note the date of his departure on the Billeting Order which he surrenders to the householder at the time of his arrival. He will further report to the office of the T. M. for clearance, bringing with him a written statement from the billet owner as to the condition in which the billet was left.

18. Persons requiring family quarters or other real property will make a written application to the T. M., accompanying such application with their authority for so doing. Upon approval of such an application, the T. M. will supply the needs by requesting the requisition of the necessary property by the German authorities. Such requests for requisition will be submitted for the approval of these Headquarters before becoming effective.

19. All requests for the requisition of real property, effective within the American Area, will be executed by the T. M. of the community in which they are to be served, and will bear the approval of these Headquarters, regardless of for whose use the requisitioned property is intended. Upon the approval of a requisition, the T. M. will issue an Assignment Notice to the applicant. This notice will give the location of the property and the amount of space assigned, and will contain a complete statement of all other conditions covered in the requisition. A copy of this Assignment Notice will be served on the property owner.

20. The needs in real property of American Troops stationed in the Allied Zones of Occupied Germany will be supplied by the commanding officers of such troops, after consultation with the proper officials of the Allied Armies in whose Area the troops are located, in accordance with the regulations in force in those Zones.

21. Billets and real property will not be obtained in the American Occupied Area except as provided above.

For whom obtained.

22. Billets will be furnished and the necessary real property will be requisitioned for all Allied Troops stationed or staging in the American Occupied Area, and to all civilians on duty in the American Occupied Area with either the American

Forces in Germany, the I.A.R.H.C. or affiliated Missions and Commissions, as may be approved by the Commanding General, American Forces in Germany.

Payment.

23. All billets and requisitioned real property furnished in the American Area will be paid for by the proper German authorities under supervision of these Headquarters.

Accomodation Allowance.

24. Accomodations provided for the housing of troops and organizations will be governed in extent by the needs of those organizations, as may be determined by these Headquarters.

25. For the housing of those persons entitled to private billets, or family quarters, suitable accomodations in accordance with the number of persons to be billeted will be provided. The following scale will be followed as far as possible; additional accomodations being assigned when necessary upon approval of these Headquarters.

Married Officers	Personal rooms	Servant's rooms	Kitchen or use of same
2nd Lieut. or 1st Lieut.	2	1	1
Captain	3	1	1
Major	4	1	1
Lieut. Colonel	5	1	1
Colonel	6	2	1

Brig. Gen. or Major Gen. (Special - as decided by Commanding General)

Extra: One or two children - one extra bedroom; three or four children - two extra bedrooms; other authorized permanent members of the family - one extra bedroom each.

Single Officers	Personal rooms	Servant's rooms	Kitchen or use of same
Lieut. or Captain	1		
Field Officers	2		
General Officers	(as ordered)		

The above to be furnished, heated and lighted, and use allowed of bathroom, if available, with hot and cold water.

N.C.O's of Grades 1, 2 and 3, when necessary in case of exceptional circumstances:

	Personal rooms	Servant's rooms	Kitchen or use of same
Grades 1, 2 and 3	2	1	1

Extra: One or two children - one bedroom; three to five children two bedrooms; six or more children - three bedrooms.

The above to be furnished, heated and lighted, and use allowed of bathroom, if available, with hot and cold water.

26. a) Proprietor's Obligations:

The proprietor is required to put the premises reasonably in repair and so maintain them during the occupation. He will be responsible for the maintenance and upkeep of the premises.

b) Occupant's Obligations:

(1) No occupant will cause any repairs or alterations to be done on the premises of which he has the use without the previous sanction of the proprietor, or in any way make any changes which would affect the number of disposition of the rooms. In case the proprietor refuses to sanction necessary repairs, the matter should be referred to these Headquarters.

(2) Change of billet will not be allowed without the previous sanction of the T. M. of the community, and this will only be granted where it can be done without damage to the billeting capacity of the neighborhood.

(3) In the event of the death of an officer, the local American authority may, at its discretion, sanction the retention of the billet by his widow or his dependents for a reasonable period.

c) Obligations of both parties:

(1) When quarters are requisitioned for the permanent occupancy of American officers and others, the owner of the quarters will be required to submit to the local T. M., in triplicate, a list of all movable furniture and equipment turned over with the quarters. The officer or person who occupied the quarters will be required to check this list and certify to its correctness on its face. A copy of the list, with the officer's certification thereon, will be filed with the requisition for the billet in the R. R. & C. Office. The officer occupying the quarters will retain one list in his possession until he vacates the premises, at which time he will make a final check of the articles in the quarters, and turn this list in to the T. M., who will forward said list to the R. R. & C. Officer, noting thereon the condition, and whether or not any of the articles are missing. No occupant will be allowed to leave the American Area before having obtained a clearance from the local American authority.

(2) Damage due to fair wear and tear shall be charged to the proprietor. Damage in excess of fair wear and tear shall be settled privately between occupant and proprietor. In the event of their being unable to agree, the matter should be referred to the local American authority for settlement.

(3) Details. (a) Where the proprietor or his servants are residents in the house, all costs of emptyings, sweeping of chimneys, heating, lighting, water, insurance rates, maintenance, upkeep and all other charges arising from the use of the building, are payable by him. Where the house is empty, the municipal authorities will pay these charges themselves or arrange with the proprietor to do so. (b) The term "The use of all furniture, fixtures, equipment and appurtenances" includes the use of a reasonable amount of glass, crockery, cutlery, table linen,bed linen and towels. All breakages and

all damage to such articles to be paid for by the occupier. (c) The term "Maintenance and upkeep" includes the repair and renovation of all billeting space, which is covered by a requisition and its equipment and furnishings. The term does not include cleaning, scrubbing, sweeping, etc. The occupants of family quarters are responsible for supplying all labor necessary in the operation of those quarters, such as personal and house service - except that which is specifically covered in the requisition. The occupants of single billets - whether in hotels or private dwellings - are entitled to all services normally furnished in a hotel. (d) It has been ruled that where an officer and his family are quartered in a house, part of which is occupied by the owner, the owner and the officer will have joint use of the grounds surrounding the house. Responsibility for the upkeep of the grounds will rest with the owner, who will make all expenditures to this end. When an officer and his family are quartered in a house which has been vacated by the owner, the officer will have the exclusive use of the grounds surrounding the house. In such cases the responsibility for the maintenances and upkeep of the grounds - such as repairs to walks, cutting of lawn and general upkeep - rests on the owner. The planting of new flowers and the embellishment of the garden, if desired by the officer, will be paid for by him.

(4) Should the furnishings and equipment of requisitioned property not be sufficient to meet the use to which such property is to be put, additional furniture and equipment will be requisitioned from the German authorities by the local T. M., with the approval of these Headquarters, upon application by the occupant of those premises. The amount and extent of the furnishings of properties will be determined by the Authorized Furniture, Furnishings and Equipment List issued by these Headquarters. Supplementary inventories for equipment so furnished by the German authorities to complete the furnishing of billets or requisitioned property will be verified and signed by the billet occupant, who thereby becomes responsible for those articles under the same condition as for the original equipment of the property.

Billets. Method of Payment.

27. At the end of each calendar month, or at the end of the period of occupancy in the case of a shorter stay, each T. M. will prepare a billeting distribution list showing the total number of billets for officers, enlisted men, animals, etc., furnished by the town, and the amount due for each during the period covered by these certificates. (See Model of B.D.L., Appendix).

28. Billeting distribution lists for quarters, offices, storerooms, kitchens, etc., furnished for the A. F. G., will be prepared in quadruplicate. B.D.L.'s for the I.A.R.H.C. and all other Allied Missions and Commissions will be prepared in quintuplicate. Each copy will be signed by the T. M. and the burgermeister, and stamped by the latter with the seal of the

town. In exceptional cases, where troops are billeted in a town and there is no T. M. present, the B.D.L. will be sent to the Z. M. of that Brigade or Division Area for signature.

Distribution of B.D.L.'s will be as follows:

Original	To the R.R.& C.Officer, A.F.G.
Duplicate	To the R.R.& C.Officer, A.F.G.
Triplicate	To Town Major's File.
Quadruplicate	To the Burgermeister of community.
Quintuplicate	To the R.R.& C.Officer, A.F.G., who will forward same to the I.A.R.H.C. or other Allied Mission or Commission concerned.

(The duplicate copy of B.D.L. will be forwarded to the German Billeting Commission by the R.R.& C.Officer, A.F.G. This commission will make payment to the burgermeister of the community in which the services were rendered).

29. The name of the householder, location of the house, together with the number of rooms and the period during which they were occupied as messes, kitchens, orderly rooms, company, detachment, supply and guard rooms will be shown on the B.D.L. Any supplies or services received, which it has been the practice of the T. M.'s to requisition in the past (not paid for by the troops), will also be shown on the B.D.L.'s.

30. Separate B.D.L.'s will be prepared for each town, village or gemeinde in which troops are billeted.

31. Messes, kitchens, orderly rooms, company, detachment, supply and guard rooms will be paid for on B.D.L.'s at rates prescribed in paragraphs 32 and 33, these regulations.

Prices for Billets.

32. The price to be paid for billets, etc., is decided by the German authorities, subject to the approval of these Headquarters. The following prices are now in force:

		Mks.	Pfgs.
a)	For each officer provided with bed, per night	6	00
	1 Living room (for Field Officer), per night	4	00
b)	Each N.C.O. or soldier, with bed, per night	6	00
	For each soldier provided with shelter only	1	00
c)	Light	1	50
	Heat	8	00
d)	For each animal in good, closed accomodation with mangers and racks, per night	0	10
	(If mangers and racks are not provided, the animals will be considered as picketed, even though under shelter, and no payment will be allowed).		
e)	Office rooms, orderly rooms, supply rooms, etc., which are located in furnished rooms and can be		

considered as officer's billets may be paid for at the rate of six (6.00) marks per day. This may be paid for on B.D.L. with proper notation after entry.

f) Land carried on B.D.L. will be paid for at the following rates:

Meadow land	5 pfgs. per sq. m. per month
Cultivated land	7 pfgs. per sq. m. per month

33. Paragraph 35, G.O. 18, G.H.Q., A.E.F., 1918, provides that it is advisable that prices for messes, kitchens, etc., be arrangedby agreement, but if no agreement can be reached the accomodation can be requisitioned. The indemnity due in this case can be fixed according to its real value. For this reason, and because there are such variations in the location and nature of such accomodations, no absolute fixed rate can be made. In view of the foregoing, it is believed that where agreement has been reached, payment can be made on B.D.L. by noting after the entry "as per agreement". Where no agreement has been made these accomodations will have to be settled according to the provisions of G.O. 104, 1920, Headquarters, A.F.G.; that is, the property will be requisitioned and the proper rental fixed by the Boards appointed for this purpose.

34. In cases where it is obvious that these rates will not furnish fair compensation, as where the whole or part of a modern office building has been requisitioned, the payment should be determined as provided in G.O. 15, 1919 and G.O. 104, 1920, Hq. A.F.G.

HEADQUARTERS
AMERICAN FORCES IN GERMANY
COBLENZ, GERMANY

19 August 1921.

Changes)
No. 1)

INSTRUCTIONS FOR THE R.R.& C. SERVICE.

Paragraphs 1, 2 and 3, Section III (Requisitions), Instructions for the R. R. and C. Service, 1 May 1921, are rescinded and the following substituted therefor:

III. REQUISITIONS.

Requisition for Supplies and Material.

The procurement of supplies and material by requisition on the German authorities will be governed by the following:

1. Local commanders are authorized without prior approval of

higher authority to enforce requisitions on the local German authorities for fuel only to meet requirements for authorized issues within regulation allowance. This authority they do not have, if such supplies are maintained from established sources of supply or from collective requisitions which may be demanded by these headquarters from German authorities.

2. Commanding Officers of towns other than Greater Coblenz (Coblenz, Pfaffendorf, Ehrenbreitstein, Coblenz-Lutzel, Coblenz-Neuendorf, Metternich) are authorized to approve and require the German authorities to fill requisitions for emergency repairs. Emergency repairs will be considered only those which, if not attended to immediately, will cause interruption to the proper functioning of their forces. In all cases where such emergency repairs are ordered by the local commanders, same will be covered by a proper requisition and forwarded to these headquarters approved and stating that the filling of the requisition has been ordered.

3. The procurement of all other classes of authorized supplies required locally will follow the rules established in these regulations. No human food, or forage, will be requisitioned without the prior approval of these headquarters, and then only in quantities sufficient to meet an emergency.

Quantities of fuel and forage when necessary to be requisitioned from the German authorities will be expressed in KILOS on all requisitions. For all other classes of supplies the ordinary U.S. classification and description will be employed.

Requisitions on the German authorities for real property (buildings, grounds, etc.), repairs and alterations to same, furniture, fixtures, and material for same, will not be included in the same requisition with other articles.

Requisitions of any nature will be approved by the Commanding Officer of the post, station or town, before being forwarded to these headquarters.

By command of Major General ALLEN:

W. H. HAY,
Chief of Staff.

OFFICIAL:
S. G. TALBOTT,
Adjutant.

Requisition of Services.

4. Local commanders are not authorized to enforce requisitions for services without the approval of these Headquarters.

5. It is contrary to the announced policy of our Government, to make direct requisition for labor except in emergency. No direct requisition, therefore, will be made for labor without the prior

approval of these Headquarters. In the event of such approval the method of payment will be prescribed.

6. When Public Buildings and Utilities, such as Government-owned buildings, barracks, depots, hospitals, water-works, electric plants, etc., which already have a working force, such as janitors, charwomen, engineers, mechanics, etc., are taken over for use by the American Forces under a German requisition, the requisition will include the ordinary and necessary personnel for its operation. The proper German official will be required to pay these employees out of public German funds at the same rate as they were paid prior to our occupation. It is incumbent upon all U.S. officers in charge of these facilities to see that these payments are promptly made.

7. When Private Buildings and Utilities, such as private-owned hotels, private hospitals, factories, etc., which already have a working force, are taken over, by a German requisition and when it is necessary and has been authorized to retain such a force, the following procedure will be observed: Individuals included in this working force will be paid at the same rate as previous to our occupation, from funds turned over by the German Government. In estimating the monthly rental on these buildings, cost of services will be considered and will be included in the total rental price recommended. The requisition will be so worded as to show this arrangement.

8. In all cases where supplies or services, as distinguished from real property, are requisitioned, requisition blank (Form 1, model in Appendix) must be properly executed, in duplicate, and forwarded direct to these Headquarters. These requisitions must be so drawn as to permit the verbage on the face of the requisition being used as the complete voucher upon which subsequent settlement may be made.

Requisitions for the use of real property.

9. The German Government, through the German Billeting Commission, will requisition the use of all real property formerly requisitioned by the R.R.& C. Service in the American Occupied Area. In order to have real property requisitioned the following procedure will be observed:

a) The service or party interested, in conjunction with an R.R.& C. representative, will locate the property.

b) The R.R.& C. representative will forward a letter in quadruplicate (Model in Appendix) to the R.R.& C. Officer, American Forces in Germany, requesting that the property be requisitioned.

c) The R.R.& C. Officer will have the letter requesting the requisition approved and the file number entered. After approval the original will be filed in the R.R.& C. Office;

the duplicate and triplicate will be forwarded to the German Billeting Commission by indorsement, and the quadruplicate will be returned to the R.R.& C. representative who originated the letter. Upon receipt of this copy the R.R.& C. representative will notify the interested service that the request for requisition has been approved and that the premises may be occupied; he will also notify the R.R.& C. Officer of the date of occupancy.

d) The rental to be paid will be determined by the German Billeting Commission, with the approval of the Army Appraisal Board, A.F.G. Payment of rentals will also be arranged for by the Germans Billeting Commission.

10. Request for requisition of real property to be used by the I.A.R.H.C., all other Allied Missions and Commissions and welfare organizations, will be forwarded in quintuplicate. The first four copies will be disposed of in the same manner as outlined in sub-paragraph (c), paragraph 9 above, and the fifth copy will be forwarded by the R.R.& C. Officer to the organization interested.

11. In urgent cases, where there is every reason to believe that the request for requisition will be approved by these Headquarters, the service interested may occupy the property before notification from the local R.R.& C. representative that the request has been approved.

Authorities Competent to Approve Requisitions and fix rental prices.

12. The only authorities competent to approve requisitions are:

For Supplies and Services - A.C.of S., for Supplies, Hq. A.F.G.

For the use of real property - A.C.of S., for Operations, Hq. A.F.G.

Rental prices to be paid for requisitions for real property are determined by the German Billeting Commission with the approval of the Army Appraisal Board, American Forces in Germany.

Disposition of Approved Requisition.

13. a) For all requisitions for supplies and services (except those services included in the requisition of a building as outlined in Section III, pars. 6 & 7, these regulations), three (3) copies of Form 1 will be executed and forwarded direct to the A.C.of S., for Supplies, these Headquarters.

b) For the use of real property, four (4) copies of

Request for Requisition Form will be executed. The original will be filed and indexed in the R.R.& C. Office, Hq. A.F.G. The duplicate and triplicate will be forwarded to the German Billeting Commission. (The duplicate will be retained by them and the triplicate forwarded to the burgermeister of the community in which the property to be requisitioned is located). The quadruplicate will be forwarded by the R.R.& C. Officer to the local R.R.& C. representative, who initiated the requisition, for his file.

14. In order to reduce the claims, which no doubt will be presented upon the evacuation of quarters, requisitioned for the use of American officers, the following procedure will be followed:

When the request for the requisitioning by the Germans of any quarters for the permanent occupancy of American officers and others is approved, the owner of the quarters will be required to submit to the T. M., in triplicate, a list of all movable furniture and equipment turned over with the quarters. The officer or person who occupies the quarters will be required to check this list, certifying to its correctness upon the face. A copy of the list with the officer's certification thereon will be filed with the letter requesting the requisition of the billet in the R.R.& C. Office. The officer occupying the quarters will retain one list in his possession until he vacates the quarters, at which time he will make a final check of the articles in the quarters and turn this list in to the T. M., who will forward said list to the R.R.& C. Office, these Headquarters, noting thereon the condition and whether or not any of the articles are missing.

Cancellation of Requisitions on Real Property.

15. One of the most important provisions in the handling of German requisitions for real property requisitioned at our request, is its prompt cancellation so that payment on same can be stopped, when use of such property is no longer required. This will be done as follows:

a) The service or party vacating the premises will notify the local R.R.& C. representative to that effect. The local R.R.& C. representative will at once notify the R.R.& C. Officer, these Headquarters. The T. M., in making this report, will give the file number on the returned copy of the letter which requested that the property be requisitioned in the first place, and the date that the requisition should be cancelled. The German Billeting Commission will be notified by these Headquarters that the use of the property is no longer desired and that the requisition may be cancelled. (See Model in Appendix). Distribution of cancellation notice will be the same as for original request for requisition of the property.

b) In cases where the vacating service or party fails to notify the local T. M., the latter will, as soon as he becomes cognizant of the fact that the premises have been vacated, find out the date upon which they have been so vacated and proceed as in sub-paragraph (a) above.

c) These provisions are very important.

Requisitions for Furniture, Furnishings, Repairs, Installations, etc.

16. Requests for the requisition of furniture, furnishings, equipment, etc., as well as requests for repairs, installations, etc., will be handled in the same way as requests for the requisition of the use of real property as outlined in paragraph 13, sub-paragraph (b) above.

17. At the end of each calendar month the burgermeister of each community in which troops are stationed will make out a detailed list, in triplicate, showing articles of furniture, furnishings, equipment, etc., supplied and repairs and installations effected during that month. This list will also show the requisition number of the quarters for which articles, etc., were furnished, and the cost price of each item. These lists will be signed by the burgermeister who thereby certifies to the correctness thereof and stamped with the seal of the community. These lists will then be forwarded to the local T. M. who will countersign them. The original and duplicate will be forwarded to the R.R.& C. Officer, Hq. A.F.G., and the triplicate will be retained in the files of the T. M. (See Model of Monthly Report in Appendix).

18. Furnishing of porch furniture for married officer's quarters is disapproved.* This disapproval is not to be interpreted to mean that the use of wicker furniture, cannot be requisitioned for interior furnishing.

19. The screening of dining rooms and kitchens is approved, and will be done whenever requested by occupants. The request will be forwarded to the R.R.& C. Office for the necessary approval, the same as requests for furniture, furnishings, etc. Complete screening of houses is disapproved, except in the cases where it is recommended by a Sanitary Officer after an investigation of the premises by him.

20. Claims for loss or damage submitted by German householders will be reduced by that amount which may rightly be attributed to fair wear and tear. In compliance with instructions contained in Par. 1178, A.R., the allowance for loss or breakage of china, glassware, etc., that may be considered as ordinary wear and tear shall be 5% per quarter or 20% per annum. All loss or breakage in excess of the above shall be born by the occupant of the requisitioned quarters. Claims on other furniture or furnishings will be discounted in the amount judged by the T. M. to be correct in each individual case.

21. The use or installation of private telephones, German line, in any property requisitioned for the use of the A.F.G. (including welfare organizations, the I.A.R.H.C. and all other

Allied Missions or Commissions) must be obtained through private agreement. They can not be requisitioned from the German authorities.

22. All requests for the installation of private telephones will be submitted to the Chief Signal Officer, A.F.G., as the Germans have no authority to make installation without the permission of that office.

23. Bulletin No. 97, 1920, Hq. A.F.G., furnishes, in detail, the cost of telephone service and installation of telephones.

IV. CLAIMS

General.

1. (1) The following instructions will govern regarding the responsibility and settlement for damages caused by our troops in the Occupied Areas.

(2) Losses or damage to the property of Germany or German citizens, caused by our troops, may be divided into 3 classes:

Class "A". Those incident to our occupation, and to the necessary functioning of our Army of Occupation, and not a result of unlawful acts or negligence on the part of individuals of the American Army.

Class "B". Those which are a result of unlawful acts or negligence on the part of individuals of the American Army, and where the circumstances are such that the 105th Article of War cannot be applied.

Class "C". Those which are a result of unlawful acts or negligence on the part of individuals of the American Army, and where the circumstances are such that the 105th Article of War can be applied.

(3) Where cases arise under Class "C", the procedure directed under the 105th Article of War may be followed; it is proper to assess and collect indemnification for the damages, and to pay them to the person or persons injured.

(4) Class "C" claims that are paid and settled under the 105th Article of War will be held in the office of the Zone Major until they are actually paid and settled. The claims will then be forwarded to the R.R.& C. Officer, Headquarters, A.F.G., together with the receipt for the money signed by the claimant. In no case will a claim be forwarded only with the notation that the matter has been referred to an organization for settlement under the 105th Article of War.

(5) In all cases, therefore, in order that the individual citizen may be protected, the procedure will be as follows, viz:

Claims will be thoroughly investigated and the amount (money value) thereof will be fixed in accordance with instructions previously issued, and by agencies already established for the purpose.

The investigated claim, with the recommendation of the Zone Major, will be forwarded to the R.R.& C. Officer, Hq. A.F.G., for disposal.

In cases where the claim has been approved by the Zone Major, and the 105th Article of War is inapplicable, a notice will be included in the claim when sent to the R.R.& C. Officer, stating the claimant's name in full; complete address - including street and number, town and kreis; and the amount for which the claim has been approved.

Claims that have been disapproved by the Zone Major, will be forwarded to the R.R.& C. Officer with an indorsement on the back of each claim stating the grounds for disapproval.

No vouchers will be executed for claims which have been approved and no "Retained Copy" of voucher will be sent to the claimant.

No cisallowal notices will be made out by the Zone or Town Majors.

(6) For cases arising under Class "B", the only action that can now be taken is to make a careful investigation and keep a record of the matter, with a view to future settlement when, and if, funds become available.

Submission of Claims.

2. When claims are made for damages alleged to have been done by individuals of the American Army to real or personal property or to the person of the inhabitant, the burgermeister or gemeinde vorsteher of the town or other municipality where such damage is alleged to have occurred shall, before presenting such claims, prepare a claim blank (Schadenersatzanspruch) covering the alleged damages. The claim blanks will be furnished the burgermeisters by the Town Majors.

3. No claims will be considered for settlement unless the claim forms have been properly filled out. No claims will be entertained that have not been presented to the burgermeister within 48 hours after the damage complained of is alleged to have occurred. Claims for damage to billets must be similarly presented within 24 hours after the departure of troops. Burgermeisters, immediately upon the receipt of claims, will investigate same and present them to the T. M. within a reasonable length of time. The Z. M. of each area will determine what he deems a reasonable length of time for his area and so notify each of his Town Majors. Claims not filed within that time will not be considered. R.R.& C.

representatives will have authority to make exceptions to the above rules and may consider a claim which is clearly just, but for some good reason the claimant has neglected to file within the period of time specified. Nevertheless, when lapse of time has been so great as to render proper investigation impracticable, the claim will be declined upon this ground. The burgermeister will be instructed to notify their constituents as to the time limits within which claims may be submitted.

Classes of Claims.

4. Most of the claims will be one of the following:

a) Damage to billets
b) Damage to land
c) Damage caused by criminal acts of soldiers
d) Damage resulting from injury to persons
e) Damage caused by A.F.G. vehicles
f) Damage caused by fire

Claims for Damage to Billets.

5. This class of claims includes damage not only to the billet itself, but also to fences, outbuildings and appurtenances of the building and damage to, or the taking of, the contents of the billet whether caused by deliberate acts, negligence, accident or unintentional injury.

6. Damage done in rooms occupied by officers and others in which they have not been regularly billeted, but which they occupied under private agreement, is not payable out of public funds.

Damage to Land.

7. Liability exists for damage to land as in other cases. This damage is generally the result of the use of land for maneuvering, training or the passage of troops. This damage can be reduced to a minimum by using training fields in areas regularly taken over or assigned for this purpose.

Criminal Acts of Soldiers.

8. In general, no liability against the United States exists in these cases. For example, losses due to thefts not occuring in billets are not compensated for out of public funds. Exceptionally, a criminal act may involve liability. This is the case where it could have been prevented by the exercise of ordinary care by superior authority. This ruling was based on the fault of responsible officers in permitting such conditions to arise. R. R. & C. representatives will not approve claims for money,

watches, jewelry and the like, alleged to have been stolen by American soldiers in billets and otherwise. Such cases should be referred to the proper officer of the troops for disciplinary action and proceedings under the 105th Article of War.

Claims resulting from injury of persons.

9. Personal injury claims will be classified, thoroughly investigated and forwarded to the R.R.& C. Officer with a complete report of the case and recommendations for approval or disapproval.

Damages caused by A.F.G. Vehicles.

10. Existing orders require that an investigation of claims for damage caused by A.F.G. motor vehicles be made by the commanding officer or an officer designated by him. In addition to this the local representative of the R.R.& C. Service will make a supplementary investigation in order to fix the responsibility for the accident, and to show whether it was due to negligence on the part of the American driver or whether it was unavoidable.

Claims for Damages caused by Fire.

11. These claims are usually for large amounts and should be carefully investigated. The method of investigation outlines in paragraph 16, under the sub-head "Investigations", in these regulations, should be rigidly followed. In many instances insurance contracts do not cover damages where troops are responsible for the fire. Damages by fire that are fully covered by insurance will not be allowed. The amount of the insurance plus the amount of the claim must not exceed the actual damage sustained by the claimant.

Investigations.

12. In every case where a claim is submitted it should be accompanied by a report of the investigation made by the burgermeister.

13. The Town Major upon receiving a claim from the burgermeister in the form as required above, shall fully investigate the alleged damage in such claim, and shall thereupon forward a report of such investigation together with the claim blank and all other papers pertaining to the case, to the Zone Major of the area in which the T. M. is located. He will keep a short record of all claims so forwarded.

14. Every report on a claim should contain at least the following points:

a) A brief statement of the circumstances under which the alleged damage was sustained, with an indication of the source from which the officer obtained the information on which the statement is based. Written statements signed by witnesses should be attached when possible. When a claim is allowed, there should also be a definite finding, based upon sufficient evidence, that the damage was caused by the American Military forces, and the unit responsible should be named when possible.

b) A statement of the damage, indicating the amount of the property damaged and the value thereof, sufficiently itemized to permit of an audit and to convey a clear understanding as to how the total was arrived at.

c) It is not intended that long reports, especially on small cases should be made. Brevity is desirable, but every report should, in as brief and concise a manner as possible, conform to the foregoing instructions. Definite recommendation, based upon the facts obtaired in the investigation, is what is wanted.

15. Reports of investigations of personal injuries should contain clear statements on the following points:

a) Whether the injured person or a member, or members, of the American Forces were entirely at fault, and if not, the degree of culpability on one side or the other.

b) In cases of injuries resulting from explosions, information should be given as to the unit concerned and the precautions that had been taken for the safety of the public. Attention should be especially directed toward determining whether the explosive was of American pattern; when accident was caused by grenades, bombs, etc., pieces of same should be obtained and forwarded when possible.

16. All reports of investigations of fires should contain the following:

a) Date of investigation.

b) Description of property, with plan of building, dimensions drawn to scale.

c) The Extent of American occupation, whether sole or partial. The parts occupied by troops immediately before the fire and the parts occupied by civilians.

d) The place where the fire started.

e) Story of the fire with sworn statements of witnesses.

f) Record of previous investigations if any have been made.

g) Facts in regard to insurance. Was claimant insured?

In what company and for how much? What action has the company taken?

h) Amount claimed.

i) Recommendations, including exact amount to be paid and person to whom payment should be made if allowance is advised, and reason against allowance if rejection is advised.

17. In addition to the investigation conducted by the burgermeister and the T. M., the Z. M. shall, if he deems it advisable, conduct a supplementary investigation. If satisfied that the claim is without merit because the facts alleged do not make a valid claim under the law, it will be the duty of the Z. M. to disallow the claim.

18. No compensation will be allowed the claimant for damages to property which may arise from or grow out of fair wear and tear. Such claims will, however, after investigation, be submitted to the R.R.& C. Officer, Hq. A.F.G. for record, in the same manner as other disallowed claims.

19. Zone Majors will classify all claims in accordance with Section IV, paragraph 1, sub-paragraph 2, these regulations, before forwarding to the R.R.& C. Officer, these Headquarters.

20. If investigation of a claim shows that any of the supporting evidence is false, or that the claimant, a board of experts, or any other individual testifying as to the validity of the claim, has intentionally mis-stated or exaggerated the facts, the individual or group of individuals responsible will be brought before a Provost Court for trial by the Z. M. The Zone Major will not forward the claim to the R.R.& C. Officer until the trial has been completed. In each case of this nature the result of the trial will be stated on the claim.

21. The R.R.& C. Officer, these Headquarters, has on file in his office, a complete file of automobile accident reports and reports of personal injury, where examination of injured person has been made by an officer of the Medical Corps, these Forces. This file is available to all Zone and Town Majors, upon request, to aid them in the investigation and settlement of all claims that may arise as the result of such accident.

<u>Claims under the 105th Article of War.</u>

22. If investigation discloses that an individual or unit may be shown to be liable under the 105th Article of War, the facts must be laid before the proper commanding officer with a view to instituting proceedings to determine the question of such liability and to order stoppage against the pay of any individual or organization found to be responsible.

23. Adjustment by stoppage of pay under the 105th Article of War, depends upon whether or not the damage was due to a wrongful act on the part of the troops. Unnecessary injury to a billet should be settled by stoppage of pay whenever practicable. On the other hand it happens not infrequently that the injury could not have been reasonably avoided. There will also be cases where, owing to lack of evidence identifying the individual or unit responsible, proceedings under the 105th Article of War will not be justified but in which it would be proper for the United States to assume responsibility for the damage caused.

24. Certain claims for damages cannot be settled under the 105th Article of War. This article, for example, covers only the taking or injury of property. Compensation cannot be made under it for injuries to persons. Nor is such method of settlement possible unless there has been some wrongful act or omission.

25. Before recommending a claim for payment the R.R.& C. representative should always satisfy himself that the case is not one which can be settled under the 105th Article of War. In every case where the R.R.& C. representative forwards a claim to a troop commander for settlement under the 105th Article of War, he should request that he be advised of the action taken so that the disposition of every claim which has arisen in his area will be known to him. Claims which are paid and settled under the 105th Article of War will be forwarded, with the receipt for the money signed by the claimant, to the R.R.& C. Officer, Hq. A.F.G., for file and record.

Settlement of Claims.

26. Zone Majors and Town Majors are not authorized to take action on claims involving more than 2,500 marks. Claims in excess of 2,500 marks will be thoroughly investigated and a report, together with a recommendation, will be forwarded with the claim to the R.R.& C. Officer. If a claim amounted to over 2,500 marks and has been reduced by the Z. M. to an amount less than 2,500 marks, and this amount is accepted by the claimant, the Z. M. will act on this claim as on other claims amounting to less than 2,500 marks.

27. If a claim is capable of settlement within the Z. M.'s or T. M's jurisdictional limitation and is not in one of the excepted classes, he will arrange adjustment promptly, and where investigation has been completed and amount to be awarded determined, will forward all papers in the case to the R.R.& C. Officer.

28. Claimants who have submitted claims which are considered to be just, and on account of and incident to the occupation by the American Army, will be so notified by these Headquarters and told to look to the German Government for settlement. The notice will be made out in triplicate and distributed as follows:

Original to the claimant.
Duplicate to the President of the Reichsvermogensverwaltung, Coblenz.
Triplicate to file in R.R.& C. Office.

All papers concerning claims will be filed, with the claims, in the office of the R.R.& C. Officer, Headquarters, A.F.G.

Important.

29. When any unit is moved from its area, the officer from that unit who has been detailed to act as T. M. will, before leaving:

a) Make out and forward to the Z. M., billeting list for the period of occupancy from date of last payment.

b) Secure from the burgermeister a statement to the effect that there are no claims against the members of the organization which has just departed except those which have been presented, and that billeting receipts have been submitted for the total period of occupancy.

c) Investigate and properly dispose of all claims which have been submitted to him by the burgermeister.

d) Forward all records in his office to the Z. M. of his area and obtain clearance certificate from him.

e) In case he is relieved by another T. M. he will turn over to his successor all records, files and forms in his office and fully instruct him in his duties before obtaining clearance certificate from the Z. M.

30. The forms mentioned in these regulations, of which models are contained in the Appendix, may be obtained, upon request, from the R.R.& C. Officer, these Headquarters.

31. All orders, memoranda or bulletins heretofore published by these Headquarters in conflict with the present regulations are hereby revoked.

HEADQUARTERS
AMERICAN FORCES IN GERMANY
COBLENZ, GERMANY.

1 February 1922.

Changes)
No. 2)

INSTRUCTIONS FOR THE R.R.& C. SERVICE.

Pursuant to authority contained in War Department Cables W-134, paragraph 1, and W-334, paragraph 1, the following changes in "Instructions for the R.R.& C. Service, 1 May 1921" are published in the name of the Secretary of War, for the information and guidance of all concerned:

Section IV - Claims - add the following:

Claims under Act of Congress approved 18 April, 1918, which reads as follows:

"AN ACT TO GIVE INDEMNITY FOR DAMAGES CAUSED BY AMERICAN FORCES ABROAD.

BE IT ENACTED BY THE SENATE AND HOUSE OF REPRESENTATIVES OF THE UNITED STATES OF AMERICAN IN CONGRESS ASSEMBLED. That claims of inhabitants of France or of any other European country not an enemy or ally of an enemy for damages caused by American Military Forces may be presented to any officer designated by the President, and when approved by such an officer shall be paid under regulations made by the Secretary of War.

Section 2. **That** claims under this statute shall not be approved unless they would be payable according to the law or practice governing the military forces of the country in which they occur.

Section 3. That hereafter appropriations for the incidental expenses of the Quartermaster Corps shall be available for paying the claims herein described.

Section 4. That this statute does not supersede other modes of indemnity now in existence and does not diminish responsibility of any member of the military forces to the person injured or to the United States".

32. Claims which come within the provisions of the foregoing Act are divided into the following classes:

(a) Claims resulting from Acts originating prior to 10 January 1920.

(b) Claims arising in any European country with the exception of France or Italy, subsequent to 10 January 1920.

(c) Claims arising in France or Italy caused incident to the American Forces in Germany or by Acts of members of these forces.

(d) Admiralty claims.

33. R.R.& C. representatives who receive claims which come under Paragraph 32 (a) and (c), will forward them immediately to the Director, R.R.& C. Service, Headquarters, American Forces in Germany, without taking any further action thereon.

34. All claims coming under Paragraph 32 (b) and (d), will

be thoroughly investigated by the representative of the R.R.& C. Service who first receives them, who will forward a report of his investigation, together with his recommendations, through his immediate commanding officer to the Director, R.R.& C. Service, Headquarters, American Forces in Germany. No travel will be ordered or performed in the investigation of these claims without orders from Headquarters, American Forces in Germany. When travel is necessary, request for travel orders will be made by the R.R.& C. representative concerned to Headquarters, American Forces in Germany, through his immediate commanding officer.

35. Upon receipt of claims coming under Paragraph 32 (a), by the Director, R.R.& C. Service, American Forces in Germany, he will forward them immediately without further action to the Foreign Claims Section, Office of the Assistant Secretary of War, Washington, D. C.

36. Upon receipt of claims coming under Paragraph 32 (c), by the Direcotr, R.R.& C. Service, American Forces in Germany, he will forward them immediately to the Chief Requisition Officer, R.R.& C. Service, 24 rue Bayen, Paris, France, with whatever information he may have or is able to obtain on the case.

37. Upon receipt of claims coming under Paragraph 32 (d), by the Director, R.R.& C. Service, American Forces in Germany, he will make a synopsis of the investigation with his recommendations, and place all papers before the A.C.of S., G-1, to be forwarded to the Foreign Claims Section, Office of the Assistant Secretary of War, Washington, D.C., by the Commanding General, American Forces in Germany.

38. Upon receipt of claims coming under Paragraph 32 (b), by the Director, R.R.& C. Service, American Forces in Germany, he will make or have made any additional investigation necessary, prepare a synopsis of the investigation with his recommendations, and place all papers before the A.C.of S., G-1, for the action of the Commanding General. Payment of claims under Paragraph 32 (b) will be made by the Finance Department out of such funds as may have been apportioned for that purpose.

39. All instructions or orders in conflict with Paragraphs 32, 33, 34, 35, 36, 37, and 38, of these instructions, are revoked.

By command of Major General ALLEN:

OFFICIAL:
S. G. TALBOTT,
Adjutant.

W. H. HAY,
Chief of Staff.

Samples of various forms used by the R. R. and C. Service and appended to this chapter.

Prepared in Quadruplicate

BILLETING DISTRIBUTION LIST

Town Village or Gemeinde ....COBLENZ............ Kreis ....COBLENZ............

NAME OF HOUSEHOLDER AND NUMBER OF HOUSE	Number of Billeting nights				Period inclusive dates		Amount due to each individual	
	Officers	E.M. N.C.O's with beds	E.M. Soldiers without beds	Animals	From	To	Marks	Pfgs
SCHMIDT, Jakob	30	90	150	200	4/1/21	4/30/21	635	00
KERN, Wilhelm	15	45	75		4/1/21	4/30/21	307	50
	MESSES, KITCHENS, ETC.							
MULLER, Peter								
1 furnished room use as Officer's Mess								
5 marks per day					4/1/21	4/30/21	150	00
1 kitchen at 2 marks per day					4/1/21	4/30/21	60	00
GERNER, Paul								
1 room, used as storeroom, as per agreement								
2 marks per day					4/1/21	4/30/21	60	00
SEE REVERSE SIDE FOR INSTRUCTIONS								
					TOTAL		1212	50

BILLETING DISTRIBUTION LIST - CONTINUED

I certify that the above is correct, that the services have been performed as stated and payment therefor has not been made.

SIGNED :............................ SIGNED....................................
The Burgermeister

SEAL (stempel) RANK:...................................
Town Major

Unit for which certificate is rendered.......Co. "C", 8th Infantry..................

....

Distribution: Original and duplicate to R.R.& C. - Triplicate to Burgermeister
Quadruplicate to T.M. File.

INSTRUCTIONS

1. Made out in quadruplicate. Original and duplicate forwarded to R. R. & C. Office, A.F.G. Triplicate given to burgermeister. Quadruplicate retained by T.M. for file.

2. Burgermeister to sign and stamp each copy. T.M. should sign each copy.

3. Write the name of the householder in CAPITALS.

4. The dates showing the period are inclusive. Therefor if a unit, for instance, leaves on the morning of the 16th, the B.D.L. should read from the 1st to the 15th inclusive. If a unit is billeted only one day in a town. - say the 6th - the entry should be made from the 6th to the 6th, so that payment on only one day will be made.

5. In the column marked "Animals", only those animals furnished with shelter, racks, etc., should be entered.

6. Rates for large storerooms, and other unfurnished rooms, should be paid for at rate of 2 pfennigs per day or "As per agreement."

7. Land paid for on B.D.L. should be paid for at the rate of 5 pfennings per sq. meter per month meadow land and 7 pfennings per sq. meter per month for cultivated land.

8. If more than one page of this B.D.L. is required, carry forward each sub-total and on last page give grand total.

(INSTRUCTIONS FOR EXECUTION OF B.D.L.)

BILLETING ORDER "B"

Quartier-Anweisung

For
Fur ____

With ____________ On ____________ St. No.
Bei ____________ Nach ____________ Str. Nr.

From ____________ To ____________
Von ____________ bis ____________ (Date of Departure)

Signature
Unterschrift ____________

Billeting Officer,
American Forces in Germany.

IMPORTANT

Hand this Billeting Order to the householder. When leaving fill in date of departure over your signature and notify Billeting Office that the billet has been vacated.

WICHTIG

Dieses Quartierbillet ist dem Nausbesitzer zu ubergeben. Weun der Offizier abreist, soll er das Datum der Abreise uber seiner Unterschrift vermerken. Unterschriebene Quartierbillete sind innerhalb 24 Stunden Lach Freiwerden der Quartiere dem Elinquartierungsamt einsureichen.

BILLETING ORDER "A"
Quartier-Anweisung

TO BE PAID FOR BY OFFICER OR PERSON OCCUPYING THE ROOM
Vom Offizier oder dem Quartiernehmer zu bezahlen.

For ______________________________
Fur

Hotel ______________________ Room No. ______________
Zimmer

From ______________ (Date Arrival) To ______________ (Date Departure)
Von Bis

Billeting Officer,
American Forces in Germany

Signature ______________________
Unterschrift

IMPORTANT :

Payment will be made at the approved rate posted in the lobby of the hotel and in each separate room.

CLEARANCE
C E R T I F I C A T E

..................19

I,..........................Burgermeister of..................

........................Kreis of...............................

Germany, do hereby certify that there are no claims to be presented by any inhabitants of the town of............Kreis of............. Germany, against any members of the United States Army except those which have been presented to the American Town Major for the period of occupancy by American troops to and including the present date 19

I further certify that due notice has been given to the inhabitants t to file all claims within forty eight hours from the time of the origin of the claim.

Signature..........................
Burgermeister

Witnessed: (Seal)

Signature..........................

Rank and
Organization.......................

Town Major.......................Germany

B E S C H E I N I G U N G

........................19

Ich,...................Burgermeister der stadt...............

in Kreise.....................Deutschland, bescheinige hiermit, dass von keinem Einwohner der Burgermeisterei.................... Forderunger geger irgend ein Mitglied der amerikanischen Armee zu machen sind, ausser solchen, welche dem amerikanischen Ortskommandant fur die Zeit der B esatzung bei amerikanischen Truppen bis zum gegenwartigen Datum einschleissend den.........................19 vorgelegt sind.

Ich bescheinige ferner, dass die Einwohrer gebuhrend angewiesen worden sind, alle Forderungen innerhaln achtundvierzig Standen nach der enstandenen Ursache des Anspruches einzureichen.

Gezeichnet...................
Burgermeister
(Siegil)

Bezeugt bei:
Signature......................
Rank and Organization..........
Town Major............... .....Germany

ARMY OF THE UNITED STATES

Form 1.

RECEIPT FOR SUPPLIES OR SERVICES

(Not Negotiable)

Date....................19

RECEIVED from..(Individual concerned).P.O. Address(Town and Kreis)

Quantity	Unit	Articles	Condition

(This form is for SUPPLIES AND SERVICES ONLY)

(Make out 3 copies and forward same direct to G-4)

Strength of Command: Officers...........Men...........Animals.........

I CERTIFY that the above supplies or services are required for use of...

....

Signature..........................

APPROVED:

.... Rank and..........................
G-4 Organization

I CERTIFY that I have received the above supplies or services and that payment therefor has not been made.

Signature..........................

Rank and..........................
Organization

NOTE TO HOLDER: This receipt should be deposited with the Burgermeister.

(Anweisung) Diese Quittung dem Burgermeister oder Geminde Vorsteher ubergeben.

Signed and approved in duplicate

....

....

THIS CANCELS REQUISITION NO. R-

May 1, 1921.

File No. R-

FROM: TOWN MAJOR, COBLENZ GERMANY.

TO: R. R. & C. Officer, Hq. A.F.G.

SUBJECT: Requisition.

1. It is desired that the German Government requisition without delay, for the exclusive use of the A.F.G. as

Quarters for a commissioned officer and family:
The entire 1st stage, consisting of three (3) rooms, exclusive use of kitchen in basement, also joint use of bath, also quarters for one servant, at No. 45 Mainzerstrasse, Coblenz, Germany, the property of Herr Con. Guerard,

including maintenance and upkeep, the necessary equipment and the necessary amount of heat, light and water.

Occupancy dates from May 1, 1921

APPROVED: Signature of T.M.

TOWN MAJOR

A. C. of S. Operations

1st Ind.

R. R. & C. Office, Hq. A.F.G., Coblenz, To:
The President of the German Billeting Commission.

1. The above request has been approved by these headquarters and is forwarded to you for the necessary action.

R. R. & C. Officer, A.F.G.

MONTHLY REPORT for THE TOWN OF....................................

for the Month of..........April 1921...................

REQUISITION NUMBER	DESCRIPTION	AMOUNT MARKS
R-1672	1 Bed -frame	750.00
	1 Spring	240.00
	1 Mattressprotector	42.00
	1 Mattress with head piece	850.00
	4 Sheets	780.00
	1 Pillow	190.00
	2 Pillow-slips	110.00
	1 Comforter	400.00
	1 Plumeau	550.00
	2 Plumeau covers	280.00
	1 Night stand	300.00
	1 Rug, 2x3 m.	1350.00
	1 Washbowl & pitcher	112.00
	1 Wash stand with mirror	2100.00
	1 Wardrobe, 2 doors with mirror	2150.00
	1 Table, small	475.00
	1 Arm Chair	550.00
	2 Bed frames) 2 Night stands).......... 2 Wash stands with mirror) 2 Wardrobes with mirror)	10600.00
	2 Springs	480.00
	2 Matress protectors	84.00
	2 Tables, oak, small	950.00
	2 Arm chairs	1100.00
	2 Mattresses with head piece	1700.00
	8 Sheets	1560.00
	2 Pillows	380.00
	4 Pillow covers	220.00
	2 Plumeaus	1100.00
	4 Plumeau covers	560.00
	2 Rugs, (1350-1550)	2900.00
	2 Washbowls & Pitcher complete	224.00
	2 Comforters	800.00
R-1699	1 Stove	493.00
R-1930	1 Stove	384.00
Ahrweiler	6 Cups and saucers	51.00
7 Kaiserstr	1 Sugar bowl	21.00
	1 Milk pitcher	13.00
	1 Tea pot	30.00
	1 Tray	35.00
	6 Spoons, tea	45.00
	6 Plates, small	25.50
von Braun-muhl, 19 Hin-denourgstr	1 Rug	3975.00

REQUISITION NUMBER	DESCRIPTION	AMOUNT MARKS
R-1933	1 Kitchen range, emanel stove pipe 2 Elbows 1 Ring	1775.00 65.00 29.00 3.00

40831.50

I certify that the above is the total amount paid by me for requisitions mentioned for the month of......April........1921

CORRECT:

(Signed)

.... Town Major

.... Burgermeister

CHAPTER NO. 11

DISPOSITION

OF

SURPLUS STOCKS, ANIMALS AND ENEMY WAR

MATERIAL.

Liquidation Commission; Sales Plans; Small Sales; Motor Transportation; Other Stocks; Animals; Enemy War Material; Miscellaneous; Officers.

DISPOSITION
OF
SURPLUS STOCKS, ANIMALS AND ENEMY WAR MATERIAL[1]

When the Third Army was sent to participate in the occupation of the Rhineland, large quantities of supplies and equipment were shipped to the American area. Some of these stocks were shipped from the United States, but the great majority came from the A.E.F. depots in France. The disbanding of the Third Army and the return to the United States of all its units having rendered these stocks surplus, it became necessary to dispose of them. This duty devolved upon the War Department Liquidation Commission which was created in February 1919.

Because of the shortage of railway transportation and of the fact that the disposition of the stocks already in France had been found to be a difficult problem, it was decided to sell the Third Army stocks in Germany. Early in July 1919 a branch office of the Liquidation Commission was established in Coblenz, and Mr. J. G. Adams was placed in charge. It was determined that if practicable the stocks should be disposed of under arrangements that would provide for payments in American dollars. Negotiations were instituted with some of the strongest German financial houses which had connections in the United States, while at the same time the Liquidation Commission, through the War Department endeavored to arouse the interest of the American correspondents of those firms. In addition, negotiations were opened with various American, French, British, Belgian, Dutch and Swedish financial interests.

[1] The Final Report of the United States Liquidation Commission (War Department 1920) forms the basis of this chapter.

The purpose of the Coblenz office of the Liquidation Commission was to arouse interest to such an extent that groups or syndicates, with ability to make payment in dollars or in dollar credits, would be formed. But while many representatives came and inspected the stocks, no offer to purchase stocks on a dollar basis was made. There were many conferences and discussions, but the representatives averred that under all the conditions, particularly with regard to the fluctuating rates of exchange, it was quite impossible to engage to make payments in dollars.

The plan to secure payments in dollars was then abandoned, and more liberal terms were offered. Notices sent to prospective purchasers stated that the stocks, as a whole or in large blocks, might be paid for in dollars, Swiss francs, Dutch guilders, British pounds, French francs of German marks, at the respective current rates of exchange; preference would be given to bids offered in those currencies in the order named. No satisfactory bid was received. Several speculators made bids for small lots of the most desirable stocks, but as an acceptance of those bids would have endangered the sale of the stocks as a whole, all were rejected.

The stocks were inventoried in dollars at cost plus 20% for transportation and handling, except motor transportation which was carried at cost plus 45%. Many commodities comprised in the stocks were also produced in Germany and could be bought there at prices that were much lower than those shown in the inventories. The depreciation of the currencies of the European countries that had been engaged in the World War, particularly the low exchange value of the German mark, was a serious obstacle to a reasonably satisfactory sale. The local purchasing power of the mark was always much higher than its exchange value in American currency, a condition that made it impracticable for Germans, or others who might plan to re-sell the stocks in Germany, to pay for them in dollars.

An effort was made to bring about the organization of a syndicate which would take the stocks in exchange for certain German products, potash for example, which could be utilized to advantage in the United States. This effort was unsuccessful as the Germans would dispose of those products for dollars only, that being the sole method by which the dollar exchange value of the mark could be strengthened.

While these efforts to dispose of all the stocks were being made, certain small sales of salvaged stocks were made. Such of the surplus Signal material as was designed for field use and had small commercial value was returned to the United States; the remainder, which was largely unserviceable in a military sense, was sold for 1,200,000 marks, worth at that time about $58,136.00. The surplus tobacco, candy and confectionery were deteriorating and had to be sold. The tobacco brought cost plus 6%. The candy and confectionery were given as inventory value of $1,200,000. It was sold early in August for 87-1/2% of the inventory valuation, which amounted to more than its original cost to the United States.

MOTOR TRANSPORTATION

In October 1919, after many attempts to dispose of the motor transportation, the Coblenz branch of the Liquidation Commission succeeded in selling it to Motor Organizations Limited, a British Syndicate at the head of which were Sir Percival Perry and Lieutenant Colonel G. W. Parkinson of the British Army.

The surplus motor transportation not only was the largest item of the surplus stocks, but also it was rapidly deteriorating from exposure to the elements, and its early sale, was a matter of much importance. After a series of negotiations the British Syndicate purchased the surplus motor equipment under the following conditions:

> The estimated purchase price was 3,250,000 pounds sterling, of which 325,000 pounds sterling was to be paid in cash.
>
> The estimated purchase price would be subject to increase or decrease upon a check of the equipment.
>
> In case of a controversy as to quantity or classification it was to be settled by the Coblenz representative of the Liquidation Commission (Lieut. Colonel T. H. Kruttschnitt, General Sales Agent, A.F.G., had replaced Mr. Adams) acting for the United States, and Sir Percival Perry or Lieut. Colonel Parkinson, acting for the Syndicate. In the event of a further disagreement, the decision of Lieut. Colonel Kruttschnitt would be final.
>
> The balance of the purchase price was to be paid in three equal installments, the first on July 1, 1920, and the last on April 1, 1921; interest at the rate of 5% per year was to be paid semi-annually from April 1, 1920.
>
> The United States was to exercise the right of supervision and to remain in constructive possession of the property, retaining a lien upon it, until the full purchase price had been paid. None of the equipment could be withdrawn and sold without the consent of the American representative and the payment to the United

States of the full purchase price of the property withdrawn.

250,000 pounds sterling was to be deposited to the credit of the American representative, as trustee, with the Paris branch of the Guaranty Trust Co. The funds were to be subject to the joint check of the American representative and the Syndicate, and were to be used to place the equipment in good condition. The funds, by additional deposits, were to be maintained at substantially the amount of the original deposit. If the American representative should become dissatisfied with the progress of the work in placing the equipment in good condition, the entire deposit might be applied toward the payment of the total purchase price.

The contract was to be construed according to the laws in effect in Washington, D.C. The British Consul General at New York was to be appointed the agent of the Syndicate; service on him could be had in any suit brought by the United States to enforce the contract.

It was believed that the interests of the United States were amply protected by the terms of this contract.

The surplus transportation was parked at several places along the Rhine and Mosel rivers, the guards being furnished by the Americans. The large and well equipped motor repair shops left by the Third Army were turned over to the Syndicate for use in placing the equipment in marketable condition. In January 1920 the great inundation of the Rhine and the Mosel caused much damage to the motor equipment; in many cases passenger vehicles were completely covered with water, while the motor trucks were in water that reached nearly to the tops of their canvas covers. After considerable discussion as to which party should bear the loss, a compromise was reached. All vehicles concerned were reclassified and payment was made in accordance with the new classification.

It was estimated that the United States suffered considerable financial loss because of the damage done to the surplus motor transportation by the floods.

OTHER STOCKS

After many conferences with representatives of the Polish Government, the remainder of the surplus stocks (except animals) was sold to the Polish Relief and Supply Corporation for 70% of the inventory valuation. These stocks comprised food, clothing, ordnance and medical supplies and a variety of miscellaneous equipment, the total purchase price being $5,536,867.71. It was understood that the Polish Government had given notes for this amount, but whether it was ever actually paid was not known by the Headquarters of the A.F.G.

A N I M A L S

In the spring of 1919, the early disbandment of the Third Army having been foreseen, boards were appointed in the various corps and divisions to weed out the unserviceable animals. These animals were then sold locally to Germans, the sales, collection and disposition of funds being conducted by the Division Quartermasters, or their authorized representatives. As the Third Army units departed for the United States their best animals were transferred to other units, and the least desirable were turned into the remount depots. During May and June 1919, 17,252 animals were sold to the French Government, which shipped them to the devastated districts in France.

On July 2, 1919, the date of the official dissolution of

the Third Army, there were on hand 29,928 animals. About the middle of the month 5,000 were sold to the Polish Government at $280 a head, payment therefor being made in Paris. These animals were shipped to Warsaw through Germany.

Early in August 4,500 animals were sold to France. The original offer had been 1,500 francs per head for horses and 950 francs for mules. In the beginning it was supposed to be a sale to France, but it transpired that the French planned to re-sell the animals to Czecho-Slovakia. When the offer of the French was rejected, their plan became known. The Americans finally sold the animals to the French for 1,800 francs, which was the price already received from Czecho-Slovakia by the French Government, plus 8% commission for transacting the business.

As soon as the units to remain in Germany had been designated, the best animals were issued to them, and orders were received to retain animals for a force that was being organized in the United States for service in Upper Silesia.

500 animals were sold to the Polish Typhus Relief Expedition. Animals that had been rejected by the French were sold in small lots to Germans. With the exception of the animals sold to the French for delivery to Czecho-Slovakia, the prices obtained by the sales to Germans were greater in most cases than those paid for the best stock by the French.

The Liquidation Commission (Judge Parker) in Paris announced in August that it had been satisfied relative to Polish credit, and that all the animals in the Wengerohr remount depot had been sold to Poland. There was some delay in obtaining the necessary cars, but finally one shipment of 400 mules was made. The next day the contract was broken; the reason therefor was never known by the American Headquarters in Coblenz.

During September 868 animals were sold at auction in small lots, and most of the remaining surplus animals were sold to a Bavarian Syndicate at 4800 marks (about $218.00) per head. The market price became quite low, but by the end of November all surplus animals remaining had been sold at an average of 3960 marks per head.

In December 1921, because of the reduction of the A.F.G., a small surplus occured and 100 animals, at an average price of 18,000 marks (then about $100.00), were sold at public auction.

ENEMY WAR MATERIAL

A large number of german motor trucks had been alloted to the A.E.F. under the terms of the Armistice Agreement. These were delivered to the Third Army after its arrival in the Rhinelands. They had been constructed during the war and had steel tires. They had undergone hard usage and by the time of their sale had been standing in the open for some months. They had deteriorated to such an extent that much repair work was required to make them available for commercial use. These "German armistice trucks" were sold en bloc for 13,000,000 marks, (then about $581,366.00).

MISCELLANEOUS

When an excess in any stocks held by a Supply Service occured, the fact was reported, with a recommendation that the excess be sold. When approved by the Commanding General, the excess stocks were then declared surplus and were sold by the General Sales Agent. The latter kept a classified list of prospective buyers; notices of sales were sent to them and placed in prominent newspapers throughout Germany. The most advantageous method was used; some sales were made at public auction, some under sealed bids, and others by circular proposal or contract.

All salvaged material was disposed of by the Salvage Officer; only those stocks that were serviceable and had been declared surplus were sold by the General Sales Agent.

During 1920 and 1921 the stocks disposed of by the General Sales Agent brought the following amounts:

	Dollars	Marks	French Francs
Miscellaneous supplies	17,744.98		
Animals	353,763.03	7,927,030.48	333,868.00
Enemy War Material		1,194,200.00	
TOTALS	371,508.01	9,121,230.48	333,868.00

OFFICERS

Lieutenant Colonel T. H. Kruttschnitt, who had been appointed General Sales Agent for the A.F.G., replaced Mr. J. G. Adams, as the Coblenz representative of the Liquidation Commission, in the autumn of 1919. After the surplus stocks left by the Third Army had been disposed of, the functions of the General Sales Agent were terminated, and Lieut. Col. Kruttschnitt returned to the United States.

On August 25, 1920, in accordance with orders from the War

Department, Lieutenant Colonel H. H. Sheen, Q. M. C., was appointed General Sales Agent, in which capacity he served throughout 1921. Major F. C. Rogers, Q. M. C., was his assistant until July 1921, when the latter was replaced by Captain E. M. Scott, Q. M. C.

CHAPTER NO. 12

DISPOSITION OF ENEMY AMMUNITION.

Ammunition Found by Third American Army; German Shells, Fuses and Powder shipped to U.S. A. for Experiments; Destruction of Large Quantities; Sale of Ammunition to German Concerns; Methods of Breaking Down; Ammunition Found by German Population; Officers.

DISPOSITION OF ENEMY AMMUNITION

When the Third American Army arrived in that portion of unoccupied Germany which came to be known as the American Area, it found large quantities of ammunition which had been left behind by the German troops when they evacuated the Rhineland. Most of this ammunition was at Trier, Neuwied, Mulheim (near Coblenz) - where the Germans had maintained a plant for assembling ammunition of various calibers, and at the Coblenz forts (Franz, Constantine, Alexander and Ehrenbreitstein).[1] The stocks comprised ammunition of German make and much that had been captured from the armies of the Allies. There were shells of every caliber and description, fuses, shrapnel, bombs, grenades, empty shell cases, and small arms ammunition, as well as quantities of material used in the manufacture of ammunition, such as powders, zinc, lead, brass, etc. Much of the ammunition was unserviceable and too dangerous to justify long storage or shipment to the United States, and plans for its early destruction or dismantling were made. However, it proved possible to ship to the United States, for experimental purposes and to form a historical collection, the following:

- 135,000 Artillery projectiles.
- 22,000,000 Rounds of fixed ammunition (2 cm.).
- 400,000 Fuses for artillery projectiles.
- 3,000 Heavy artillery cartridge cases (brass).
- 36,000 Naval shells (17 cm.).
- 2,000 Tons of German powders.

[1] Ammunition found in outlying districts was collected at railheads and shipped by rail to Mulheim.

Large quantities of the remainder were completely destroyed in demolition pits in the vicinity of Trier, in the woods between Bassenheim and Ochtendung, and in a clay quarry near Karlich.

7,700 tons of the ammunition were sold in June 1919 to the South German Diskonto Company of Mannheim. The sale was made under a contract which provided that the ammunition should be converted into scrap and that the base materials could be sold for commercial uses only. The contract was let for 1,298,900 marks, which at that time amounted to something over $100,000. 5,979 tons were sold in August 1919 to Doctor Heinrich Ahrens of Wiesbaden. The terms of this contract were similar to those of the other, and the consideration was 2,414,767 marks, the exchange value of which was between $175,000 and $200,000. As guarantees for the completion of the contracts, the American military authorities required a deposit of 300,000 marks from the South German Diskonto Company and one of 500,000 marks from Doctor Ahrens.

For the breaking down of this material three plants were maintained - one at Hallschlag for artillery ammunition, one at Rubenacher Schanzo for small arms ammunition, and one at Neuwied for grenades. The actual work was done by the contractors, under the direct supervision of military inspectors.

Powder was disintegrated in water or burned. The explosive charges were removed from grenades and then destroyed, after which the bodies were broken up or melted. Small arms ammunition was reduced to its component parts; the powder was destroyed and the metal salvaged and sold. The methods of breaking down artillery ammunition were briefly as follows:

Removal of cartridge case.

Destruction of propelling charge.

Firing of primer.

Destruction of case.

Removal and destruction of fuse.

Removal of rotating band from shell.

Removal and destruction of bursting charge.

Removal of base charge from shrapnel.

Removal of shrapnel head.

Removal of balls and matrix.

Destruction of shell bodies by breaking up or melting.

The work on the first contract was completed in February 1922, and that on the second contract in November 1921.

From time to time the German inhabitants reported the finding of stray shells, grenades, etc., in the American area. Military personnel were promptly sent to the places whence the reports emanated, and the ammunition was destroyed at once.

OFFICERS

The dismantling of enemy ammunition was supervised by the Ordnance Officer. Colonel (afterwards Major) James K. Crain, Ordnance Department, served in this capacity until July, 1921, when, because of his return to the United States, he was relieved by Lieutenant Colonel Harry B. Jordan, O. D.

CHAPTER NO. 13

DISMANTLING OF FORTIFICATIONS.

Provisions of Treaty of Versailles; Dismantling Agencies; Coblenz Fortifications; The Categories; Dismantling Operations; The New Categories; The American Policy; Summary; Officers; Policies of the Allies.

DISMANTLING OF FORTIFICATIONS.[1]

PROVISIONS OF TREATY OF VERSAILLES

Article 180 of the Treaty reads as follows:

> All fortified works, fortresses and field works situated in German territory to the west of a line drawn fifty kilometers to the east of the Rhine shall be disarmed and dismantled.
>
> Within a period of two months from the coming into force of the present Treaty such of the above fortified works, fortresses and field works as are situated in territory not occupied by Allied and Associated troops shall be disarmed, and within a further period of four months they shall be dismantled. Those which are situated in territory occupied by Allied and Associated troops shall be disarmed and dismantled within such periods as may be fixed by the Allied High Command.
>
> The construction of any new fortifications, whatever its nature and importance, is forbidden in the zone referred to in the first paragraph above.
>
> The system of fortified works of the southern and eastern frontiers of Germany shall be maintained in its existing state.

DISMANTLING AGENCIES.

In order to supervise the execution of those provisions of the Treaty of Versailles that prescribed the disarmament of Germany, the Inter-Allied Military Control Commission was established. Its President was the French General Nollet, who made his headquarters at Berlin. The Control Commission was under the direction of the Allied Military Committee of Versailles, at the head of which was Marshal Foch. The dismantling of German fortifications, as prescribed by Article 180, was among the duties of the Control Commission. For this purpose General Nollet established at Cologne a sub-commission of Allied officers (including Italians), which

[1] Policies are set forth at the end of this chapter.

had under its jurisdiction the dismantling of the Coblenz fortifications.

This sub-commission specified exactly what works or portions thereof were to be destroyed, and its directions were forwarded to the Germans for compliance.

In May 1920 the Germans established a Dismantling Office (Enfestingungsamt) at Coblenz. This body was charged with the preparation and execution of detailed plans for the dismantling of the fortifications in accordance with the directions received from the Allied Sub-Commission at Cologne, provided that such dismantling was authorized by the American Commanding General. After the authority of the latter had been given, the actual work was done by German contractors.

The Coblenz Dismantling Office was under the direction of Lieutenant Colonel Ruger, a retired officer of the German Army. He performed his duties in a very satisfactory manner, and the relations between his office and the Americans were cordial.

The Engineer Officer represented the American Commanding General in matters relating to fortifications. No dismantling of fortifications in the American zone could be undertaken except by authority of the latter. The Engineer Officer kept the American Headquarters informed upon everything relating to fortifications, and he maintained an office of record for all correspondence, reports, maps, etc., connected with that subject. Frequent conferences with the Sub-Commission at Cologne and with the German Dismantling Office were held. (Of the former body the two officers who came into the most intimate contact with the Engineer Officer were Colonel of Engineers, A.B. Ver Eecke of the Belgian Army, who was President of the Sub-Commission, and Major Thebes of the French Engineers.) No demolition that tended to endanger the troops or property of the United States was permitted, and frequent inspections of the progress of the dismantling and of its compliance with specifications were made.

COBLENZ FORTIFICATIONS

West (left) bank of the Rhine. The fortifications on the left bank consisted of Forts Alexander and Constantine and Feste (Fort) Eranz, the latter with several outlying works, Buberheimer Flesche (redan), Rubenacher Schanze (redoubt), Metternicher Schanze and Moselflesche. Most of these defenses had been constructed between 1817 and 1827, and all were obsolete as compared with modern works. They had not been used as forts since about 1900, but as they provided a certain amount of overhead cover (15 feet in some cases) they were still of some defensive value.

East (right) bank of the Rhine. Here the fortification system consisted of Fortress Ehrenbreitstein and its smaller outlying works - Fort Rheineck, Pleitenberg Work and Arzheimer Schanze, and of Fort Asterstein, which included Fort Rheinhell (with Glockenberg Work and Bienhorn Schanze). Fortress Ehrenbreitstein was of obsolete type, but because of its natural strength and underground defenses it was still of considerable value as a strong point in opposing a river crossing at Coblenz. The other works (built between 1860 and 1870) were obsolete also, although they too provided a certain amount of overhead cover. About 1914 the fortifications system on the right bank had been strengthened by the construction of twenty-eight concrete shelters.

There were no other fortifications in the American area.

THE CATEGORIES

In the spring of 1920, in a letter from General Nollet to Lieutenant General von Cramon, the German Disarmament Representative at Berlin, the German fortifications subject to dismantling were divided into three categories -

Category No. 1

Works to be temporarily preserved.

Category No. 2

Works to be partially dismantled at once, the remaining parts to be temporarily preserved.

Category No. 3

Works to be totally and immediately dismantled. This category included all works not coming under Category No. 1 or No. 2.

In accordance with the instructions contained in the letter of General Nollet, the fortifications in the American zone were classified as follows:

Category No. 1

Left bank - Forts Alexander and Constantine and Schanze Grossfurst Alexander.
Right bank - all works.

Category No. 2

Left bank - Feste Franz and outlying works (Bubenheimer Flesche and Rubenacher Schanze).
Right bank - none.

Category No. 3.

Left bank - Metternicher Schanze and remains of Mosel-flesche.

DISMANTLING OPERATIONS

During the years 1920 and 1921 the following fortifications (all on the left bank) were dismantled wholly or in part:

Feste Franz (only partially)
Bubenheimer Flesche
Rubenacher Schanze
Metternicher Schanze

On the right bank the only work destroyed was Bienhorn Schanze.

As the work at Moselflesche was but a stone flanking tower and was used by the American troops, its destruction was not permitted.

Feste Franz was the only work of importance. It was a masonry and earth fort in the north-west section of Coblenz-Lutzel (on the left bank of the Mosel river, opposite Coblenz), which had been built during the period 1816-22. Connected to it were two redoubts, Bubenheimer Flesche and Rubenacher Schanze; the former was constructed of masonry and earth, while the latter, except for a small powder magazine, consisted of earthworks only. Metternicher Schanze, which was near Feste Franz, was a small earth redoubt and contained a masonry powder magazine.

The dismantling of Bubenheimer Flesche was commenced on July 20, 1920, and by December 8 had been completed. The underground galleries were blown up, the masonry walls and parts of the earther parapet and rampart were destroyed, and the moat was filled in. Between October 10 and November 12, 1920, the powder magazines at Rubenacher Schanze were demolished; the earth works, being of no importance, were left intact.

The work on the demolition of Feste Franz was commenced on November 6, 1920, and was completed on August 10, 1921. The underground galleries were blown up or walled in, the walls, store-rooms and other masonry construction were blown up, the earthen parapet and ramparts were levelled, and the moat was filled in. An average of fifty men were employed in this demolition work, and about 16,000 pounds of explosive (dynamite and powder) were used.

Bienhorn Schanze, a small earthen work near Pfaffendorf on the right bank, was destroyed by levelling the parapet and filling in the moat. This was done between August and October 1921.

THE NEW CATEGORIES

In September 1921 the fortifications in the occupied territories were re-classified as follows:

1st Category

Works whose destruction might be undertaken at once.

2nd Category

Works whose destruction had to be temporarily postponed.

Under the new instructions works in the first category were to be destroyed by the Germans as before, while the others were to be destroyed by the armies of occupation before their withdrawal from Germany.

The American Commanding General then decided that no more works in the American zone should be destroyed at once, and so placed all remaining fortifications in the second category. A month later, in the nature of a compromise with the insistent French, he agreed to permit Schanze Grossfurst Alexander, the powder magazine at Feste Franz, and Moselflesche to be placed in the first category.

THE AMERICAN POLICY

The policy of the French, who generally succeeded in having their way in matters pertaining to the disarmament of Germany, was without delay to dismantle as many fortifications as possible. This became the policy of the Sub-Commission at Cologne, but it was not that of the American Headquarters at Coblenz. The policy of the latter was to proceed slowly, and where fortifications were obsolete or were used by the American troops, to disapprove projects for their immediate demolition.

The Germans, naturally enough, were not desirous of early dismantling. There was a sentimental reason for disliking to see old and historic defenses dismantled, and then the expenses were an additional drain on finances already dangerously low. Further, there was probably a hope that where the dismantling of certain fortifications was delayed it might be possible eventually to preserve them from destruction.

SUMMARY

During 1920 and 1921 one fort and four redoubts had been destroyed, and authority had been given to destroy three smaller works. On the left bank there remained Forts Alexander and Constantine and three lesser defenses, and on the right bank Fortress Ehrenbreitstein, Fort Asterstein and various smaller works were still intact. There had been several requests that some of the remaining fortifications be placed in the first category, but at the end of 1921 there was no immediate indication that this would be done, although it might develop that no objection would be made to the dismantling of fortifications no longer needed for the accommodation of American troops, especially those on the left bank of the Rhine. It was practically a fixed policy, however, that Fortress Ehrenbreitstein, because of its historic, sentimental and scenic value, should be kept intact.

OFFICERS

The officer having supervision of the dismantling of fortifications in the American Area during the greater part of 1920 and 1921 was Major Bernard A. Miller, Corps of Engineers (deceased). After his death in December 1921 the duty was performed by Major Edwin A. Bethel, Corps of Engineers.

POLICIES OF THE ALLIES

The letter set forth below shows the policy of the Inter-Allied Military Control Commission with regard to the dismantling of fortifications in the occupied territories, from the beginning of operations until September 1921.

INTER ALLIED MILITARY CONTROL COMMISSION

Staff, 5th Section No. 248. Berlin 23 April 1920.

From: General NOLLET.

To: Lieut. General von CRAMON.

Paragraph 180 of the Treaty of Peace (2nd sub-paragraph) determines that the fortified works, fortifications and fortified places, located within the area occupied by Allied and Associated troops have to be disarmed and dismantled, that is within a period, determined by the Allied commanders.

I have the honor of informing you that the dismantling of these works have to be performed according to the following orders:

Category No. 1. Works to be temporarily preserved. The dismantling of these will be postponed to a date which will be determined later. (See enclosed list No. 1).

Category No. 2. Works of which partial dismantling is ordered immediately. The parts which have to be preserved temporarily will be dismantled under the same conditions as the works of Category No. 1 and further orders will be issued. (See enclosed list No. 2).

Category No. 3. Works which have to be dismantled totally and immediately. They include all works (without any exceptions) not falling under Category No. 1 or No. 2.

I beg you to begin with the performance of this dismantling at once. This work has to be performed totally by the German Government, the temporarily postponed as well as the works which have to be performed immediately, and is superintended by the District Commission in Cologne for the cities of Mainz, Coblenz, Coln (Cologne) and Nezel, and by the Commission in Stuttgart for the cities Kehl and Germersheim.

With regard to the question raised in a letter No. F 37/20 of April 17th of this year, we inform you that it falls under general orders and I have directed the District Commission of Cologne to get in touch with the American Army of Occupation.

(Signed) NOLLET.

LIST NO. 1.

INTER-ALLIED MILITARY CONTROL COMMISSION.

STAFF 5TH SECTION.

List of works located in the Occupied Territory which have to be preserved temporarily.

(Paragraph 180 of the Treaty).

CITIES	NAMES OF THE WORKS	REMARKS
Coblenz	Bridgehead works on the right bank Fort Alexander, left bank Fort Constantine, left bank	These works will be dismantled at a time not yet determined, the time will however be determined by the Allied authorities.

LIST NO. 2

INTER-ALLIED MILITARY CONTROL COMMISSION.

STAFF 5th SECTION.

List of the fortified works located in the Occupied Territory which have to be partially and immediately dismantled.
(Paragraph 180 of the Treaty)

CITIES	NAMES OF THE WORKS	REMARKS
Coblenz	Right bank - none. Feste Franz - left bank and works connected; Bubenheimer Flesche, Rubenacher Schanze	Only those buildings and storages which are used by troops will be preserved. All other parts have to be dismantled. The preserved parts will be destroyed later on under the same conditions as those named in List No. 1.

Due to the advanced stage of the dismantling of the fortifications in Category No. 3, it was deemed necessary to re-classify the defensive works into two new catogories, as explained in the following letter:

COMMAND OF THE ALLIED FORCES
OF OCCUPATION IN THE RHINE-
LAND TERRITORIES.
GENERAL STAFF
POSTAL SECTOR 77.
INTER-ALLIED BUREAU OF OPERATIONS.
No. 1407.

G.H.Q. 29 September 1921.

From: General Degoutte, Commanding the Allied Forces of Occupation in the Rhineland Territories.

To: Commanding General, A.F. in G., Coblence.

By letter 1659 of the 13th July 1921 General NOLLET, President of the Inter-Allied Military Commission of Control informed the Marshal President of the Allied Military Committee at Versailles that the advanced state of the labors in the fortification work to be destroyed and to be dismantled immediately, permitted the assumption that this task would be completed in the period fixed, i.e., before the 10th December 1921.

On this date there will only remain for the integral execution of Article 181 of the Treaty of Peace the destruction of the works situated in Occupied Territories and provisionally retained for the convenience of the Armies of Occupation or for reasons of a military nature.

Having regard to the difficulties which we shall experience in having the destruction of these works carried out by the Germans themselves when we evacuate the Rhineland Territories and at the moment when the Inter-Allied Military Commission of Control will have probably ceased to operate, the Marshal President of the Allied Military Committee at Versailes has decided that each one of them will be subject to a new classification into one of the following categories:

1st Category	Works whose destruction may be undertaken at once.
2nd Category	Works whose destruction has had to be temporarily posponed.

The Works of destruction of the 1st Category will be executed in the same conditions as those effected up to the present.

The execution of the works of the 2nd Category is to be provided for by the troops of occupation themselves before their withdrawal and under the responsibility of the Generals Commanding in Chief the Armies of Occupation.

Plans of destruction established by the Sub-Commission of the Fortifications indicate for each work the destructions to be executed and the time necessary for the execution of the work.

I beg to forward you herewith in communication the plans relative to the town of Coblence where all works have been maintained up to date with the exception of Fort Franz which has been dismantled. The works which remain to be executed are therefore, as General NOLLET observes in the letter attached No. 6884 of the 23rd September 1921 to Marshal FOCH, extremely important.

I should be very obliged if you would kindly examine, being guided by the above considerations, if it is possible to diminish the number of works to be maintained, while taking account of the needs of the defense and the necessities of billeting.

In the affirmative, may I request you to let me know as soon as possible what are the works whose maintenance is in your opinion indispensable, with, if possible, statement of the time during which they should be retained.

for Chief of Staff,

(Signed) P. HERING.

CHAPTER NO. 14

INTER-ALLIED COMMISSIONS.

Inter-Allied Railway Commission - Organization, Duties, Reserve Coal Stocks, Strikes, Officers; Inter-Allied Waterways Commission - Organization, Duties, Patrols, American Sector, Control Posts, Strikes, Officers; D.G.C.R.A. - Communications, Supply, Liaison, Miscellaneous, Officers.

REFERENCE.

General . . . . . A. M. G., Volume I, Chapter 3.

INTER - ALLIED COMMISSIONS[1]

INTER-ALLIED RAILWAY COMMISSION[2]

This Commission consisted of a delegate from each army of occupation, and in the beginning of 1920 had its seat at Trier. Its president was the French General Payot, who held the title of Director General of the Communications and Supply of the Armies, and whose main office was at Wiesbaden. The Commission, under the Armistice Agreement, remained under the orders of the Commanding General of the Allied Armies of Occupation and constituted a part of the Staff of the Allied High Command. It was divided into various sub-commissions; an officer of the Army that occupied a particular territory, or the larger portion therof, was at the head of the delegation which represented the Commission in that area. The American Delegate on the Commission was under the supervision and control of G-4 who had a liaison officer with General Payot at Wiesbaden. Railway matters in the American zone were under the Coblenz Delegation, at the head of which was an American officer. The German Government had a representative with the Commission and with each sub-commission or delegation; his duties were to receive the orders of the body to which he was attached and to transmit them to the proper railway Direktion.

The seat of the Railway Commission was moved to Wiesbaden in March 1920. The liaison officer with General Payot, in addition to his other duties, was then appointed American Delegate with the Commission. The A.F.G. maintained a representative with the

[1] For Inter-Allied Rhineland High Commission, see Volume I, Chapter.

[2] A. M. G., Volumes I (Page 76) and IV (Page 1278).

sub-commission at Cologne and, for a time, with the sub-commission at Trier.

After all transportation activities had been placed under the Quartermaster (September 1920), the Railway Section of the Transportation Service was consolidated with the American Section of the Railway Commission. Thereafter the President of the Coblenz Delegation reported to the American Delegate with the Commission and to the Transportation Officer on matters concerning the transportation of the A.F.G.

The permanent status of the railways in the occupied territories was established by Ordinance No. 6[1] of the High Commission, which was published in general orders to the A.F.G. on January 22, 1920. No material change in the duties of the Railway Commission was effected, and they remained practically the same as they had been on January 1. In general these duties consisted in the exercise of control in the following particulars:

> Maintenance by Germany, under the Armistice Agreement, of a stipulated amount of rolling stock at pre-Armistice standards.
>
> New Construction.
>
> Movement of war material.
>
> Movement of Allied and "Armistice" rolling stock, the passage of which into unoccupied Germany was forbidden.
>
> Coal Stocks.
>
> Statistics of traffic between occupied and unoccupied Germany.
>
> Transportation for the Armies of Occupation.

[1] A. M. G., Volume IV, Page 1316.

Procurement of data for use in preparing plans of defense.

The control was exercised through observers and control posts; the latter were located at important points on the perimeters of the bridgeheads and at such other places as was necessary for the maintenance of an effective supervision over traffic.

The maintenance of sufficient reserve stocks of coal for the railways was a difficult problem, generally because of strikes in the coal districts and occasionally because of natural causes such as floods.[1] At the beginning of 1920 there were sufficient reserves, the various yards having on hand an average of 30 days supply. The January inundations caused the stocks to be greatly reduced, some of the larger terminals having as little as two or three days' supply. Ordinance No. 6 required a minimum reserve of 15 days' supply at each terminal. This was found inadequate, and Ordinance No. 60, which rescinded so much of Ordinance No. 6 as related to coal reserves, vested in the High Commission the authority to prescribe the minimum amount of coal to be kept on hand. On December 4, 1920, the High Commission, by an executive order, fixed the minimum coal reserves at 35 days' supply. In spite of this precaution the coal stocks were reduced to 18 days' supply in December 1921, because of labor troubles in the coal districts and a railroad strike.

While railway strikes and other labor troubles in unoccupied Germany frequently affected the coal supply of the occupied territories, there were no serious railway strikes in the Rhineland

[1] Volume II, Chapter 6 and 22.

until December 1921.[1]

The Rhineland Agreement[2] provided in Article 6 that services might be requisitioned by the Armies of Occupation, and in Article 10 that railway personnel should obey any orders given, "for military purposes", by the Commander-in-Chief of the Allied Armies or his representatives. Article 13 authorized the High Commission, in its discretion, to declare a state of siege. Ordinance No. 6 prescribed that upon the declaration of a state of siege, the Railway Commission should take over the management of the railways. Ordinance No. 53 provided for arbitration of labor disputes by a Board of Conciliation appointed by the High Commission, and prohibited the inauguration of strikes until after such attempts to settle disputes had been made.[3]

In December 1921 the German labor unions ordered a general strike on all railways, including those in the unoccupied territories.[4] In compliance with these orders a strike was called on the lines of the Cologne Direktion, under the jurisdiction of which were the lines, on both banks of the Rhine, between Coblenz and Cologne. The strike order provided specifically that sufficient train service for military communications and for the transportation of military supplies should be maintained in the occupied territories. It was claimed by the labor unions that this proviso would serve to fulfill all obligations under the

1 Up to this time the American area had been practically free from even minor railway strikes.

2 A. M. G., Volume IV, Page 1381.

3 Ordinances and orders of the High Commission were transmitted to the Railway Commission and similar bodies through the Headquarters of the Allied Armies of Occupation.

4 Volume II, Chapters 1 and 2.

Rhineland Agreement.[1]

Immediately after the publication of the strike order, General Payot, in the name of the Railway Commission but without securing the consent of the entire body, directed the sub-commission at Cologne to requisition the services of the railway employees "under Article 13 of the Rhineland Agreement". The Cologne sub-commission, in the name of the Commander-in-Chief of the Allied Armies of Occupation, then issued orders requisitioning the entire personnel of the Cologne Direktion; the authority under which the services were requisitioned was quoted later as Article 6 and 10, instead of Article 13, of the Rhineland Agreement. Orders were issued to the German delegate with the sub-commission, directing that the normal train service be maintained. The American Commanding General declined to permit the posting of the requisition notices in the American area, stating that they could not be made effective there without his consent.[2] Later the American Headquarters issued orders forbidding the strike and directing striking employees to return to work.

The strike order was not generally obeyed in the American area. At first only the shop, yard and maintenance of way laborers ceased work. The train crews remained at their post, but later, under the pressure of the striking employees, a number of them

1 The occupying authorities maintained throughout that any interruption of traffic was a menace to the safety of the forces of occupation.

2 The American Headquarters feared that if the French once secured possession, by requisition, of the railways in the areas occupied by them, it would prove exceedingly difficult to induce them to return the railways to German management; in addition it was not believed that a requisition of services was necessary.

struck. The railways managed to maintain an irregular train service, but the strike was unquestionably effective, although a rather large minority of the employees declined to participate. After a few days the strike was called off. Some of the strikers in the occupied territory had obeyed the military orders and returned to work at once, but the majority did not return until so ordered by the leaders of their unions.[1]

Due to the conflicting policies of the various authorities of occupation it could not be said that the situation caused by the railway strike had been well handled. At the end of 1921 no fixod policy for the solution of similar problems had been adopted.

OFFICERS

Col. J. H. Elliott, T. C., was the American Delegate from August to November 1919, when he was relieved by Major G. A. Holmes, T. S., who served as such until February 1920. Major G. G. Bartlett, Inf., was then appointed American Delegate, continuing in that capacity throughout 1920 and 1921.

Major Holmes served as President of the Coblenz Delegation in addition to his duties as American Delegate. In the former capacity he was replaced in November 1919 by Captain C. D. Symes, T.S., who, in turn, was relieved in August 1920 by Captain W. P. Morse, Inf. The latter was replaced in September 1920 by Captain W. H. Sadler, Q.M.C., who remained in that position throughout 1921.

[1] The strike had been initiated by radicals who had secured control of the central unions. The great majority of the men in the American area had no desire to strike.

INTER-ALLIED WATERWAYS COMMISSION.[1]

On January 1, 1920, the organization of the Waterways Commission was similar to that of the Railway Commission, while its seat was at Cologne. Its President was General Payot, it remained under the orders of the General Headquarters of the Allied Armies of Occupation, and it constituted a part of the staff of the Allied High Command. The American Delegate was stationed in Coblenz, attending only the important meetings of the Commission; a subordinate at Cologne handled routine matters.

The Rhine was divided into four sectors, one of which was controlled by each army of occupation. In addition to the preparation of plans of incorporation into the general scheme of defense, the duties of the Commission were -

Control of Navigation.

Supervision of the river transport of the forces of occupation.

Transmission of orders to the German authorities and to river personnel.

Police of the Rhine.

Movement of war material.

Statistics of traffic and cargoes.

Ordnance No. 18 of the High Commission, which placed river transportation upon a permanent status, was made effective in the American area on April 21, 1920. Since the beginning of the year the Waterways Commission had exercised only such control as was necessary to insure the safety and communications of the armies of occupation; hence the provisions of Ordnance No. 18 made no change

[1] A.M.G., Volume I (Page 79) and IV (Page 1326).

in the functions of the Commission. No important changes in the organization or duties of the Commission occured during the period 1920-21.

Orders concerning river nagivation were transmitted to the Schiffahrtsabteilung, Gruppe West.[1] Through this organization the necessary water transportation for the supply of the forces of occupation was secured by requisition, as were patrol boats for policing the river, landing stages, floating cranes, etc.

As the German river police had not been permitted, since the Armistice, to carry firearms, the Waterways Commission policed the river with patrol boats which worked in conjunction with German shore police. Despite the precautionary measures used, thefts of coal from barges assumed large proportions from time to time. The patrol boats were also used for the inspection of shipping, cargoes, manifests, etc.

During the coal shortage that occured in January 1920, barges containing about 13,000 tons of coal were seized by the American Delegation and held as a reserve for the A.F.G. and public utilities in the American area. 2600 tons of this coal were unloaded and used; the balance was released as soon as navigation had become normal and the coal stocks had been increased.

The original sector placed under the jurisdiction of the

[1] The Schiffahrtsabteilung was organized during the war for the management of navigable waters and the co-ordination of shipping activities. It was a section of the Ministry of Public Works. At the request of the Allied authorities it had continued to function in the occupied territories.

American Delegation had extended from Rolandseck on the north to Bacharach on the south, a distance of over 65 miles. In March 1921, at the request of the American Headquarters, the control of that portion of the sector lying south of Rhens was turned over to the French; this in order that the limits of the sector under American control might coincide with the limits of the American area. The sector under the control of the American Delegation was thus reduced to a length of approximately 45 miles.

The central control post was at Coblenz, another post being maintained at Remagen near the northern limit of the sector. In August 1920 a post was established at Duisburg in the neutral zone; this was done in order to insure the prompt passage of barages containingsupplies for the A.F.G. This post was removed later to Homburg (across the river in occupied territory). Through the posts at Remagen and Homburg (and the Cologne representative) close touch was maintained with American supplies during the voyage from the Belgium-Holland border to the American area.

During the floods of January 1920 there was a complete stoppage of river traffic for approximately three weeks. A large number of drums of gasoline and oil were set adrift and flowed downstream; the patrol boats of the Waterways Commission, together with many small motor boats and barges, were used in recovering the drums.

From August to December 1921 navigation was greatly hampered by the unusually low stage of the Rhine. Many tugs and barges went aground; as barges could carry but very light cargoes, a marked increase in freight tariffs was permitted.

In July 1921 the tug crews suddenly struck, and many barges were tied up in the American area. Extra patrol boats were requisitioned and put into use. The patrols compelled tugs to keep moving through the area. thus frustrating attempts to desert, and at the same time clearing the American sector of accumulated shipping. The strike was called off some ten days later, an agreement between owners and crews having been reached. This was the only river strike that occured in the American zone between the beginning of the occupation and the end of 1921.

OFFICERS

Captain G.W. L. Prettyman, Cavalry, served as American Delegate from January 1920 until January 1921, when he was relieved by Major J.W.H. Reisinger, Infantry. The latter served in that capactiy until January 3, 1922, when he was replaced by Major G.G. Bartlett, Infantry.

Captain T.C. McCormick, F.A., served throughout 1920 and 1921 as the American representative at Cologne.

D. G. C. R. A.[1]

The D.G.C.R.A was formed on November 19, 1918, by an order of Marshal Foch. It formed a part of his Inter-Allied Staff, and at the end of 1921 it had the same status, maintaining a liaison officer at his headquarters in Paris. It was moved to Wiesbaden in September 1919, and on January 10, 1920 (the date upon which the Treaty of Versailles became effective), it was placed at the disposal of the Commander-in-Chief of the Allied Armies of Occupation. Each of the four armies of occupation had a representative with the D.G.C.R.A.: these officers and their attached personnel composed the Inter-allied body of that name.

[1]Direction General des Communications et Ravitaillements aux Armees-Director General of the Communications and Supply of the Armies. This was actually an inter-allied body, the foregoing being the title of the officer at its head. The body was generally termed the D.G.C.R.A.

The various delegations were called missions. The American Mission with the D.G.C.R.A originally consisted of several officers and a number of field clerks and soldiers; after the latter part of 1920 but one officer, with a greatly reduced personnel, remained on this duty. In addition, he served as the American member of the Inter-Allied Railway Commission.

As its name implied, the D.G.C.R.A. had two main functions - matters connected with Communications and those relating to Supply. It functioned also as an organ of liaison for various matters. Its duties, under the three headings, were as follows:

Communications:
- Rail - through the Railway Commission.
- Water- through the Waterways Commission.
- Highways, through the Roads Commission.[2]
- Wire and Radio, through the Inter-Allied Telegraph Bureau.[3]

Supply:
- Food, forage and fuel.

Liaison:
- Plan of defense.
- Armistice material
- Destruction of permanent facilities for German mobilization
- Accumulation and distribution of various kinds of information.
- Questions concerning communications and supply during the World War.
- Communications and supply of allied forces in plebiscite districts.

---

[2] Volume II, Chapter 6. This body consisted of representatives from the four armies, and met at irregular intervals to coordinate matters relating to highways; it was not permanent in the same sense as the Railway and Waterways Commissions.

[3] This had an organization similar to that of the Roads Commission, and lik the latter was not a continuing body, meeting only when matters relating to wire or radio communications required coordination.

Through the D.G.C.R.A the various matters were brought up, coordinated and agreed upon; in this way conflicting interests, if any, were harmonized and mutual cooperation was secured. Questions were usually placed before the D. G. C. R. A. in the form of letters or memoranda from the armies concerned; if the question concerned only one army, the decision was made by the D.G.C.R.A., otherwise copies were sent to the other armies through their representatives and the question was placed on the agenda of the mext conference.

Conferences were held, usually once a month, at Wiesbaden. In addition to the various representatives with the D.G.C.R.A., these conferences were attended by representatives of indicated Staff Divisions and Services of the four armies, and by representatives of the Headquarters of the Allied Armies of Occupation, the French Ministry of War, and of the Inter-Allied Railway and Waterways Commissions. From time to time a representative of the Inter-Allied Rhineland High Commission would attend a conference. In these conferences general discussions of the questions on the agenda were had and decisions were arrived at if practicable; when necessary, the conference made recommendations to the armies or to the High Commission as to the action to be taken upon certain questions.

Communications.

Ordnance No. 6 of the High Commission, which placed the railways in the occupied territories upon a permanent status, was drawn up after exhaustive discussions in conferences of the D.G.C.R.A. After it was put into effect on January 10, 1920, it was amended from time to time, but only after the armies, through the D.G.C.R.A., had agreed upon the new provisions.

The D.G.C.R.A. maintained a general supervision over railway matters, particularly with regard to strikes. When strikes occurred, notably the strike of December 1921,[1] concerted action by the various commanders and the High Commission was taken through the D.G.C.R.A.

Through the Waterways Commission the armies controlled navigation and policed the rivers. Late in 1920 the Americans proposed that the military personnel used in policing the Rhine should be reduced and that an armed German police be substituted. The views of the four armies conflicted, and not until the end of 1921 was an agreement reached; a very small body of armed German police was authorized, the armament being in accordance with suggestions put forward by the Americans.

There were frequent labor disputes in the Rhine shipping, and strikes occurred from time to time. (Strikes were not permitted in the American zone, the measures taken proving effective in every case.) The D.G.C.R.A. was kept in touch with labor conditions in the Rhine shipping, the Intelligence Services of the various armies contributing to this end. Through its acquaintance with social and economic conditions among the crews, stevedores, etc., the D.G.C.R.A. aided in securing legitimate increases in their wages, and at the same time helped to arrange compromises which served to bring strikes to an end.

The highway system in the Rhinelands was divided into zones

[1] Volume II, Chapters 1 and 2.

according to the four armies of occupation, the roads in each zone being under the sole control of the particular army; in this respect the highways differed from the railways and waterways. Each army was responsible for the maintenance of roads in its zone. One of the main functions of the Inter-Allied Road Committee was the proper distribution of road material, which was drawn from quarries in neighboring sections of unoccupied Germany as well as in the Rhinelands. The fact that three different languages (English, French and German) were spoken by those using roads in the occupied territories made necessary the consolidation and co-ordination into one code of the various regulations that had been prescribed in the four zones. After two months study by its members, assisted by the Road Commission, the D.G.C.R.A. drafted a code of road regulations for use throughout the occupied territories. This code was enbodied into a proposed ordinance and submitted to the High Commission for approval. This was done late in 1921, and it was expected that the necessary ordinance would be published early in 1922.

Supply

Up to the spring of 1920 most of the forage used by the A.F.G. had been purchased in France, in the vicinity of Bordeaux. In May of that year the French Ministry of Public Works, through the D.G.C.R.A., invited the attention of the A.F.G. to the increasing quantity of good forage obtainable in and near the occupied territory, and requested the Americans to secure their forage in Germany, thus preventing interference with forage deliveries to the French Army (in France). The Americans complied with the request.

As the prices of the food, forage and fuel secured from the Germans by requisition entered into the calculation of the costs of occupation, a check upon the amounts charged by the Germans in the various zones was necessary. The D.G.C.R.A. made this check, securing the necessary information from the various armies and furnishing them with tables showing the prices charged for similar articles in the various areas.

Matters concerning fuel stocks[1] were among the most important activities of the D.G.C.R.A. The subject of fuel was constantly studied and was discussed at every monthly conference. It was recognized that the spring and summ.r, the period before the movement of the crops, was the time when the greatest efforts to establish the coal stocks at the required standard[2] should be made; it was not always possible to accomplish this, but its attainment was the constant endeavor of the D.G.C.R.A.

Liaison

The defense scheme, or the "Plan of Defense", in so far as the important matters of Communications and Supply were concerned, was a function of the D.G.C.R.A., which prepared the "supply, communications and evacuation" annex thereto. Representatives of the D.G.C.R.A. participated in the war games[3] at the Inter-Allied General Headquarters) in which the plan of defense was used and through which modifications of the plan were evolved. The

[1] Volume II, Chapters 6 and 22.

[2] For the armies 90 days' supply, public utilities 30 days and railways 35 days.

[3] In March 1920 the war game was played in the presence of Marshal Foch and the four army commanders. This gave rise to much comment in the European press, which claimed that an advance further into Germany was imminent.

many changes in conditions (number of effectives, locations of troops, improvements of railway, etc.) made it necessary that the annex mentioned above be frequently modified in order that it might always be up to date.

The A.F.G. were not concerned with Armistice material to so great an extent as were the associated forces, but the railway control posts on the perimeter of the bridgehead aided in this respect. These posts prevented the egress of Armistice material and sent to the D.G.C.R.A. lists of such material as had been prevented from passing into unoccupied Germany.

The matter of the destruction of permanent facilities for German mobilization was handled, in large part, by the D.G.C.R.A. The decisions regarding the eventual destruction of the Mulheim ammunition depot, the Bendorf railhead, and other establishments in the American area were received from the Inter-Allied Military Control Commission and transmitted to the armies by the D. G. C. R. A. The decisions could not be made effective until the American Commanding General had approved them; hence none of the facilities mentioned had been destroyed up to the end of December 1921.

The D.G.C.R.A., through the representatives of the four armies of occupation, collected much information concerning communications, supply and related matters. This information was collected by the D.G.C.R.A. and then distributed to the Armies. The various representatives also served in an auxiliary capacity to the Military Board of Allied Supply in its preparation of a comprehensive report of its activities during the World War.

General Payot was the President of this Board and Brigadier General Charles G. Dawes was the American member.

The United States was never represented in the control of the plebiscite district in Upper Silesia. During the latter part of 1919 and the early portion of 1920, when it was still undecided whether the Provisional Infantry Brigade would go to Upper Silesia, the American Mission with the D.G.C.R.A. was active in making tentative arrangements for the movement.

Miscellaneous

In addition to the activities coming under the three classifications shown above, the D.G.C.R.A. was concerned with minor matters of various kinds. Customs exemptions for military personnel and their families traveling in Allied countries, whether on duty or on leave, were secured, and a uniform identity card in three languages was adopted.

Officers

General Payot of the French Army continued as the President of the D.G.C.R.A. from its organization throughout 1920 and 1921.

The American Mission was originally headed by Lieutenant Colonel (then Colonel) George M. Russell, G.S., who remained in that position until February 1920. He was replaced by Major (then Lieutenant Colonel) George G. Bartlett, Infantry, who was still on that duty on December 31, 1921.

CHAPTER NO. 15

THE PERMANENT INTER-ALLIED ARMISTICE COMMISSION.

The American Section; American Armistice Commission, Repatriation of German Prisoners of War; Separatists; Requisitioning Funds for A.F.G.; Officers.

REFERENCE.

General . . . . . . A. M. G., Volume I, Chapter 13.

THE PERMANENT INTER-ALLIED ARMISTICE COMMISSION

The original seat of the Permanent Inter-Allied Armistice Commission (P.I.A.A.C.) was at Spa, Belgium, where it began to function shortly after the Armistice. The American Section was composed of a number of officers and was headed by a major general, with a lieutenant colonel as Chief of Staff. On July 11, 1919, Cologne became the seat of the P.I.A.A.C., which continued to function there, although with a greatly reduced personnel, until the ratifications of the Treaty of Versailles were exchanged in January 1920.

The activities of the American Section covered almost every phase of the conditions embodied in the Armistice Agreement and the subsequent protocols and conventions, seven in all. Its major activities were -

Repatriation of Prisoners of War (American and German).
Enforcement of Armistice conditions.
Communication with the German Government relative to incidents of the occupation and to the affairs of individual Americans and Germans.

On August 30, 1919, the American Section was disbanded, and its records were forwarded to G.H.Q., A.E.F. In order to maintain liaison with the Armistice Commission and to facilitate communication with the German Government, the Commanding General, A.F.G., in September, 1919, appointed a staff officer as American representative, with station at Cologne.

After the dissolution of the Armistice Commission in January 1920, the American Section continued to function as the American Armistice Commission, and on January 24, 1920, it moved to Coblenz. It was arranged that the American Commissioner in Berlin should serve as a medium of communication with the German Government. The American Commissioner opened his Berlin office (the old American Embassy) in January 1920, and the arrangement mentioned greatly facilitated the transaction of business.

Correspondence relative to the exchange of German prisoners of war held by American troops was handled through the American Section. Beginning September 17, 1919, the German prisoners were repatriated at the rate of about 3000 men per day. In all 755 officers and 35,461 men were turned over to the German authorities at Limburg and 3,921 men at Coblenz.

The separatist movement in the Rhinelands came within the purview of the American Section, which conducted a good deal of correspondence on the subject. The American authorities took the stand that no interference with the constitutional system of government could be permitted, a far-reaching decision which probably did more than anything else to render abortive the plans of the separatists for a coup d' etat.[1]

The activities of the American Armistice Commission, as it functioned under the A.F.G., were as follows:

> Payments due by Germany to American prisoners of war.
>
> Investigation and collection of claims against the United States by German prisoners of war who had been held by American troops.
>
> Investigations concerning American prisoners of war in Germany, German prisoners held by American troops, and German soldiers reported missing in front of an American sector.
>
> Securing funds for the maintenance of the A.F.G. from the German Government.
>
> Location of graves of American prisoners of war buried in Germany, and of the graves of American prisoners and dead behind the old German lines in the theater of operations.

Cases coming under the first three groups were handled by

[1] A. M. G., Volume I, Chapter 15.

direct communication with the Prussian Heeresabwicklungsamt (Army Liquidation Office) and the Heeresfriedenkommission (Army Peace Commission), both of which operated under the Ministry of War in Berlin.

Much delay was encountered in securing funds, in accordance with the Armistice Agreement, from the German Government for the maintenance of the A.F.G. Under the original procedure the A.F.G. informed the American Armistice Commission as to the amount required for maintenance during a certain period. The Commission would then request the German Government to turn over the indicated amount. The American requisitions for funds had to pass through a number of offices before reaching the German Ministry of Finance, and usually there was a delay of fifteen days before the funds reached the A.F.G. Hence there was negotiated an agreement that the requisitions for funds might be made directly upon the Reichsvermogensverwaltung (National Property Administration) in Coblenz the latter office was authorized to make direct application to the Reichsschatzministerium (National Treasury Department) for permission to place the required amount at the disposal of the Finance Officer, A.F.G. By this arrangement the procurement of funds from the German Government was greatly facilitated.[1]

By June 1921 the activities of the American Armistice Commission had about come to an end. The Commission was not officially discontinued, but on December 31, 1921, there were no cases pending and its work, for all practical purposes, had been completed.

OFFICERS.

[1] Volume II, Chapter 9.

Major General C. D. Rhodes
(to December 31,1918)

Brigadier General Malvern Hill Barnum
(to August 30, 1919)

Lieut. Col. L. A. Craig, F. A.
(Chief of Staff)

Colonel F. S. Young, G. S.
(September 1919 to June 1921)

Major C. K. Nulsen, A. G. D.
(June to December 31, 1921).

CHAPTER NO. 16

THE AMERICAN FORCES IN GERMANY.

Creation; Command; Organization; Provisional Infantry Brigade; Air Service Detachment; Order of Battle, December 27, 1919 and January 1, 1922; Strength; Station Lists; Officers.

THE AMERICAN FORCES IN GERMANY.

CREATION.

In accordance with orders from the General Headquarters of the American Expeditionary Forces, the Third American Army passed out of existence on July 2, 1919, and the staff and troops remaining in Germany were officially designated as the American Forces in Germany (A.F.G.), effective July 3, 1919.

The blue circular patch with a superimposed red "O" containing a white "A," which had been the insignia of the Third Army (Army of Occupation), became the insignia of the A.F.G., and all officers and enlisted men were directed to wear it.

COMMAND.

Lieutenant General Hunter Liggett was in command of the Third Army on the date it was dissolved and its headquarters transferred to the S.O.S. On July 3, 1919, Major General E. F. McLachlin assumed temporary command of the A.F.G., being relieved on July 8 by Major General Henry T. Allen, who had been selected as the permanent commander.

Brigadier General Malin Craig, who had been Chief of Staff of the Third Army, departed for the United States shortly after the arrival of General Allen, and Brigadier General W.W. Harts became Chief of Staff of the A.F.G. He was ordered to the United States in April 1920, and Colonel John C. Montgomery, G.S., who had been G-3, served as Chief of Staff until November 1920, when he returned to the United States. The next Chief of Staff was Colonel W. H. Johnston, G.S., who remained in that position until May 1921, when he was appointed Brigadier General. Colonel W.H. Hay, G.S., then became Chief of Staff, and was serving as such on December 31, 1921.

ORGANIZATION.

It had been decided by the United States that its permanent force of occupation should consist of what was practically a small reinforced brigade of approximately 7000 men.

The summer and early autumn of 1919 was a period of constant troop mevements. In the latter part of August the 1st Division, the last of the combat divisions of the Third Army, departed for the United States, but it was not until October that all auxiliary units of the Third Army had left the American area.

In July the 8th Infantry had arrived to form the bulk of the A.F.G. The combat units of the permanent force were to be --

8th Infantry
2nd Battalion, 6th Field Artillery
7th Machine Gun Battalion
Provisional Cavalry Squadron (2 troops)
Co. A, 1st Engineers
1st Field Signal Battalion.

In addition to Quartermaster, Ordnance, Motor Transport, Medical and other auxiliary units there was a provisional guard battalion of 4 companies which had been sent from France to guard the large surplus stocks left by the Third Army. This battalion was disbanded in February 1921, its personnel being absorbed by other units of the A.F.G.

Troops of the 38th French Division took station in the Coblenz Bridgehead upon the departure of the 1st American Division, but remained only a few weeks, the 8th Infantry then furnishing the necessary guards on the perimeter.

Provisional Infantry Brigade. The United States Government had agreed to furnish a contingent of troops for duty in Upper Silesia during the plebiscite which was to take place there after the ratification of the Treaty of Versailles. Accordingly the Provisional Infantry Brigade, or Silesian Brigade as it was generally designated at the time, arrived in the American area in early November (1919). This brigade consisted of 2 infantry regiments (5th and 50th), a provisional machine gun battalion and some auxiliary units. It was announced that pending its departure for Upper Silesia, the brigade would be stationed in the American area and, under the control of the A.F.G., would undergo a course of training.

After the failure of the United States to ratify the Treaty, this brigade became an integral part of the A.F.G., remaining as such until the end of 1921 when it was disbanded.

While the Provisional Infantry Brigade was considered at first but a temporary addition to the A.F.G., it seemed probable that its stay in the American area would be indefinite. The combat troops of the A.F.G. were then organized into two brigades. The 8th Infantry, 2nd Battalion of the 6th Field Artillery, 7th Machine Gun Battalion and the Mounted Detachment (or Provisional Cavalry Squadron) were organized into the 1st Brigade, A.F.G., and the Provisional Infantry Brigade became the 2d Brigade, A.F.G. Brigadier Generals F. W. Sladen and W. H. Sage, respectively, were the brigade commanders. In the spring of 1920 General Sage returned to the United States and was succeeded by Colonel W. H. Johnston, Infantry, who remained in command of the 2d Brigade until the late autumn of 1920, when he became Chief of Staff of the A.F.G. Brigadier General H. C. Hale, who had come from the United States, then took command of the 2d Brigade and retained it until late in 1921 when he was promoted

Major General and ordered to the United States. Colonel H. E. Knight, 5th Infantry, was in command of the 2d Brigade from the relief of General Hale until its dissolution on December 31, 1921.

General Sladen was ordered to the United States in the early summer of 1921, and Brigadier General W. H. Johnston was then assigned to the 1st Brigade, remaining in command of it throughout the remainder of the year.

Air Service Detachment. In September 1920, an Air Service Detachment of 13 officers and 88 men arrived from the United States. It was equipped with 12 planes and took station at the Weissenthurm flying field where barracks, shops, hangars, etc., were erected for its use.

The organization of the A.F.G. remained practically the same until the end of 1921. Then in accordance with War Department orders, the 50th Infantry was placed on the inactive list and the machine gun battalions were disbanded. The 5th Infantry was then transferred to the 1st Brigade, and the 2d Battalion of the 6th Field Artillery and the Mounted Detachment were removed from the 1st Brigade and placed directly under Headquarters, A.F.G. As already stated, the 2d Brigade was disbanded on December 31, 1921.

No replacements were received after December 16, 1920. There was a steady flow of men to the United States, each transport carrying several hundred; some of these were men whose enlistments were approaching expiration, some were undesirables, othere were being sent back because of marriage, while a few were general prisoners. During the latter part of 1921 the reduction ordered by the War Department was in progress and a large number of men were sent back, while some 3000 more were to be returned early in 1922.

NOTE. In other chapters of this work descriptions of various activities of the A.F.G. will be found.

ORDER OF BATTLE.

On December 27, 1919 the "order of battle" of the A.F.G. was as follows:

HEADQUARTERS
A.F.G.

HEADQUARTERS
HEADQUARTERS TROOP

ENGINEERS
Co. "A", 1st Engineers

MOTOR TRANSPORT
Motor Command No. 41

MEDICAL
Base Hospital, A.F.G.
Medical Supply Depot, A.F.G.

ORDNANCE
Ordnance Detach., A.F.G.

QUARTERMASTER
Post Quartermaster-Coblenz
Quartermaster Depot
Q.M. Sub-Depot
Port of Antwerp

MISCELLANEOUS
Replacement Depot, A.F.G.
Prov. Military Police Co.
Military Prison Detach.
Disciplinary Barracks Detach.
Prov. Transportation Co. No. 1
Railway Trans. Detach.
A.P.O. 927
Postal Detach., P.E.S.

1ST BRIGADE
A.F.G.

HEADQUARTERS
Brigade Headquarters

CAVALRY
Mounted Detachment, A.F.G.

INFANTRY
8th Infantry Regiment
7th Machine Gun Bn.

ARTILLERY
2nd Bn., 6th Field Artillery.

2ND BRIGADE
A.F.G.

HEADQUARTERS
Brigade Headquarters

INFANTRY
5th Infantry Regiment.
50th Infantry Regiment.
Prov. Machine Gun Bn. A.F.G.

MOTOR TRANSPORT
Motor Command No. 34
Service Park Unit No. 540

MEDICAL
Prov. Sanitary Train.

ORDNANCE
Mob. Ord. Rep. Shop No. 12.

ENGINEERS
Co. "D", 1st Engineers.

SIGNAL CORPS
2nd Brigade, A.F.G., Sig. Detach.

QUARTERMASTER
Prov. Bakery Unit No. 1

UNITS ATTACHED BUT NOT FORMING PART OF THE AMERICAN FORCES IN GERMANY.

INFANTRY
Prov. Guard Bn., (ASC)

MISCELLANEOUS
Office of Civil Affairs.
American Section, Inter-Allied Ry. & Inland Waterway Comm.
American Polish Relief Exp.

On January 1, 1922, after the disbanding of the 2d Brigade, the organizations of the A.F.G. were -

HEADQUARTERS

HEADQUARTERS, A.F.G.

ENGINEERS.
Co. "A" 16th Engineer Bn.
Co. "A" 1st Engineers.
Co. "B" 1st Engineers.
Co. "D" 1st Engineers.

QUARTERMASTER
Supply Service
Coblenz
Coblenz-Lutzel
Bendorf
Andernach
Mayen
Port of Antwerp, Belgium.
Detachment Trans. Service
Motor Transport Service

ARTILLERY
2nd Battalion 6th Field Artillery.

ORDNANCE.
18th Ordnance Co.
60th Ordnance Co.
82nd Ordnance Co. (Inactive)

MEDICAL
Station Hospital
Medical Supply Depot
Prov. Medical Bn.
Station Vet. Hosp.

AIR SERVICE
Detachment

SIGNAL CORPS
20th Service Co.

CAVALRY
Prov. Cav. Squadron

HEADQUARTERS COMMANDANT

Headquarters Detachment
Postal Detachment
Provost Marshal's Department and Provost Marshal

Disciplinary Barracks
Det. 8th Inf.
Military Prison
Det. 8th Inf.

Prov. Military Police Co.

1ST BRIGADE, A.F.G.

HEADQUARTERS	INFANTRY
Brigade Headquarters	5th Infantry, Regt.
	8th Infantry, Regt.

STRENGTH

The total strength on December 31, 1919, was 842 officers and 17,986 men, of whom 115 officers and 1726 men were attached. There were also 96 field clerks and 81 nurses.

On December 31, 1920, the total strength was 622 officers and 15,887 men, of whom 32 officers and [illegible]6 men were attached. In addition there were 67 field clerks and 66 nurses.

On December 31, 1921, the total strength was 465 officers, 8245 men, 21 warrant officers, 28 field clerks and 44 nurses. 93 officers and 3028 enlisted men were in excess of the newly authorized strength and were to be returned to the United States as fast as transportation became available.

STATION LISTS.

On December 27, 1919, the troops of the A.F.G. were stationed as follows:

HEADQUARTERS AMERICAN FORCES IN GERMANY	COBLENZ
HEADQUARTERS TROOP, A.F.G.	COBLENZ
HEADQUARTERS 1ST BRIGADE, A.F.G.	COBLENZ
HEADQUARTERS 2ND BRIGADE, A.F.G.	ANDERNACH
CAVALRY	
Mounted Detachment, A.F.G.	Ehrenbreitstein
INFANTRY	
Hqrs. 5th Infantry Regiment	Andernach
Headquarters Company	Andernach
Supply Company	Andernach
Machine Gun Company	Andernach
1st Battalion	Andernach

2nd Battalion	Andernach
3d Battalion	Plaidt
Hqrs. 8th Infantry Regiment	Coblenz
Headquarters Company	Coblenz
Supply Company	Coblenz-Lutzel
Machine Gun Company	Ehrenbreitstein
1st Battalion	Coblenz
2nd Battalion Hqrs.	Neuwied
Company E	Montabaur
Detachment	Nieder Erbach
Detachment	Malmeneich
Detachment	Hundsangen
Detachment	Molsberg
Detachment	Herschbach
Detachment	Mohren
Detachment	Elbingen
Detachment	Dilkheim
Company F	
Detachment	Dierdorf
Detachment	Puderbach
Detachment	Hilgert
Detachment	Steinmel
Detachment	Hohrhausen
Detachment	Epgert
Detachment	Gullesheim
	Breitscheid
Company G	Neuwied
Detachment	Honningen
Detachment	Waldbreitbach
Detachment	Weissfeld
Detachment	Remagen
Detachment	Engers Bridge
Company H	Selters
Detachment	Muntersbach
Detachment	Wolferlingen
Detachment	Rossbach
Detachment	Obersayn
Detachment	Strassen
Detachment	Schenkelberg
3d Battalion	Coblenz
Hqrs. 50th Infantry Regiment	Mayen
Headquarters Company	Mayen
Supply Company	Mayen
Machine Gun Company	Mayen
1st Battalion	Mayen
2nd Battalion	Mayen
3d Battalion	Nieder Mendig.
7th Machine Gun Battalion	Ehrenbreitstein
Detachment	Wirges
Prov. Machine Gun Battalion, A.F.G.	Ochtendung
Prov. Guard Battalion, Hqrs.	Coblenz-Neuendorf
Companies Nos. 1 and 2	Coblenz-Lutzel
Company No. 3	Coblenz-Neuendorf
Company No. 9	Coblenz
Company No. 11	Sinzig

Detachment	Bitburg
Company No. 12	Andernach
Detachment	Mulheim
Company No. 13	Bendorf
ARTILLERY	
2nd Battalion, 6th Field Artillery	Coblenz-Lutzel
ENGINEERS	
Company "A", 1st Engineers	Coblenz
Company "D", 1st Engineers	Meisenheim
SIGNAL CORPS	
1st Field Signal Battalion	Coblenz
2nd Brigade, A.F.G., Signal Detachment	Thur
MOTOR TRANSPORT	
Motor Command No. 34	Kruft
M.T. Co. No. 405	Kruft
M.T. Co. No. 737	Kruft
Service Park Unit No. 540	Kruft
Motor Command No. 41	Coblenz
M.T. Co. No. 333	Coblenz
M.T. Co. No. 334	Coblenz
M.T. Co. No. 335	Coblenz
M.T. Co. No. 336	Coblenz
Prov. Service Park Unit No. 1	Coblenz
Prov. Service Park Unit No. 2	Coblenz
MEDICAL	
Base Hospital, A.F.G.	Coblenz
(Convalescent Camp)	
(Hospital Train No. 70)	
(Field Hospital No. 13)	
(Evac. Amb. Co. No. 26)	
Medical Supply Depot, A.F.G.	Coblenz
Prov. Sanitary Train, A.F.G.	Mayen
Field Hospital No. 6	Mayen
Field Hospital No. 8	Mayen
Ambulance Company No. 6	Mayen
ORDNANCE	
Ordnance Detachment, A.F.G.	Metternich
Mobile Ord. Repair Shop No. 12	Kruft
QUARTERMASTER	
Post Quartermaster	Coblenz
Post Q.M. Detachment	Coblenz
Quartermaster Depot	Coblenz-Lutzel
Q.M. Depot Detachment	Coblenz-Lutzel
Forage Office	Andernach
Railhead Office	Andernach
Railhead Office	Sinzig
Railhead Office	Mayen

Quartermaster Sub-Depot	Bendorf
Port of Antwerp	Antwerp, Belgium
Prov. Bakery Unit No. 1	Mayen

MISCELLANEOUS UNITS

Replacement Depot, A.F.G.	Coblenz
Prov. Military Police Company	Coblenz
Detachment	Andernach
Detachment	Wiesbaden
Detachment	Cologne
Detachment	Bonn
Detachment	Remagen
Prov. Military Prison Detachment	Coblenz
Prov. Disciplinary Barracks Detachment	Coblenz
American Sec., Inter-Allied Ry. & Inl. Wy.	Coblenz
Prov. Transportation Co. No. 1	Coblenz
Railway Transportation Detach.	Coblenz
A.P.O. 927	Coblenz
Postal Detachment, P.E.S.	Coblenz
Office of Civil Affairs	Coblenz
American Polish Relief Exp.	Warsaw, Poland

All stations in Germany, unless otherwise noted.

On January 1, 1922, the station list was as set forth below:

HEADQUARTERS	Coblenz
HEADQUARTERS COMMANDANT	Coblenz
HEADQUARTERS, 1ST BRIGADE, A.F.G.	Coblenz

CAVALRY

Prov. Cavalry Squadron	Ehrenbreitstein
Headquarters Detachment	Ehrenbreitstein
Troop "A" & "B"	Ehrenbreitstein

INFANTRY

Headquarters 5th Infantry	Andernach
Headquarters Company	Andernach
Service Company	Andernach
Howitzer Company	Andernach
1st Battalion (Less Co. "D")	Andernach
Company "D"	Engers
2nd Battalion (Less Co. "H")	Mayen
Company "H"	Engers
3rd Battalion (Less Company "M")	Asterstein
Company "M"	Engers

Headquarters 8th Infantry	Coblenz
Headquarters Company	Coblenz
Service Company	Coblenz-Lutzel
Howitzer Company	Coblenz
1st Battalion (Less Co. "D")	Coblenz

Company "D"	Ehrenbreitstein
2nd Battalion (Less Company "H")	Coblenz
Company "H"	Ehrenbreitstein
3rd Battalion (Less Company "M")	Coblenz
Company "M"	Ehrenbreitstein
Medical Detachment	Coblenz
Disciplinary Barracks Det.	Coblenz
Military Prison Detachment.	Coblenz
Headquarters Detachment	Coblenz
Postal Detachment	Coblenz
Guard Detachment	Antwerp, Belgium
Provisional Military Police Co.	
Headquarters	Coblenz
Detachment	Andernach
Detachment	Wiesbaden
Detachment	Cologne
Detachment	Mayen
Detachment	Antwerp, Belgium
Detachment	Aschen
Detachment	Bonn
Detachment	Boppard
Detachment	Bad Ems
Detachment	Niederlahnstein
Detachment	Montabaur
ARTILLERY	
2nd Battalion 6th Field Artillery	Coblenz-Lutzel
Headquarters Detachment & Combat Train	Coblenz-Lutzel
Batteries "D", "E", & "F"	Coblenz-Lutzel
ENGINEERS.	
Co. "A" 16th Engineer Battalion	
Headquarters & Supply Platoon	Coblenz
1st Platoon	Kreuzberg
2nd Platoon	Kettenheim
Co. "A" 1st Engineers	Coblenz
Co. "B" 1st Engineers	Coblenz
Do. "D" 1st Engineers	Coblenz
SIGNAL CORPS	
20th Service Company	Coblenz
QUARTERMASTER	
Supply Service	
Post Quartermaster	Coblenz
Depot Quartermaster	Coblenz-Lutzel
Detachment	Bendorf
Detachment	Andernach
Detachment	Mayen
Port of Antwerp	Antwerp, Belgium

Detachment Transportation Service	
Headquarters	Coblenz
Detachment	Honningen
Detachment	Remagen
Detachment	Elbingen
Detachment	Neidererbach
Detachment	Wiesbaden
Detachment	Cologne
Detachment	Mayen
Detachment	Andernach
Detachment	Engers
Detachment	Aachen
Detachment	Wasserbillig
Detachment	Brussels, Belgium
Detachment	Antwerp, Belgium
Detachment	Bendorf
Detachment	Puderbach
Detachment	Paris, France
Motor Transport Service	
Headquarters	Coblenz-Lutzel
M.T. Co. No. 60	Coblenz
M.T. Co. No. 63	Coblenz
M. Rep. S. No. 100	Coblenz-Lutzel
MEDICAL	
Station Hospital	Coblenz
Hospital Co. No. 60	Coblenz
Hospital Train No. 70	Coblenz
Medical Supply Depot	Coblenz
Station Veterinary Hospital	Coblenz-Lutzel
Prov. Medical Battalion	Neuwied
Ambulance Co. No. 56	Neuwied
Hospital Co. No. 57	Neuwied
Hospital Co. No. 58	Neuwied
ORDNANCE	
18th Ordnance Co.	Metternich
60th Ordnance Co.	Mulheim
82nd Ordnance Co. (inactive)	Metternich
AIR SERVICE	
Detachment	Weissenthurm
FINANCE	
Detachment	Coblenz
MISCELLANEOUS UNITS	
Civil Affairs	Coblenz
Civil Affairs	Mayen

POST OFFICES

Coblenz, Germany	A.P.O. No. 927
Antwerp, Belgium	A.P.O. No. 944

RECRUITING STATIONS AND OFFICERS

Germany - Station Hospital, Coblenz — Major F.M. Fitts, M.C.
Belgium - Hqrs., Port of Antwerp — Capt. O.L. Graham, M.C.

All stations in Germany unless otherwise noted.

OFFICERS

There follows a list of all officers on duty with Headquarters, A.F.G., between July 3, 1919, and December 31, 1921:

		FROM	TO
COMMANDING GENERAL			
Allen, Henry T.	Major Gen.	July 1919	----------
Aides to Commanding General			
Allen, Henry T. Jr.	Capt. Cav.	July 1919	March 1920
Sumner, W. B.	Capt. F.A.	Mar. 1920	----------
Hutchinson, C.G.	Capt. Cav.	June 1920	----------
Dunn, W. McK.	Capt. F.A.	Nov. 1920	----------
Barten, A.I.	2nd Lt. F.A.	July 1919	Sept. 1919
McHenry, James	1st Lt. F.A.	Sept 1919	June 1920
CHIEF OF STAFF			
Craig, Malin	Brig. Gen.	July 1919	July 1919
Harts, W.W.	Brig. Gen.	July 1919	April 1920
Montgomery, J.C.	Major G.S.		
(Col. to June 30, 1920)		Apr. 1920	Nov. 1920
Johnston, W.H.	Col., G.S.	Nov. 1920	May 1921
Hay, W.H.	Col., G.S.	May 1921	----------
Hossfeld, H. (Att.)	Lt. Col., Inf.	Dec. 1921	----------
SECRETARY GENERAL STAFF.			
Harris, Arthur	Major G. S.	July 1919	July 1919
Walton, L.B.	2nd Lt. C of I	July 1919	Apr. 1920
Rumbough, D.S.	Capt., F.A.	Apr, 1920	Jan. 1921
A. C. OF S. G-1			
Reynolds, S.C.	Lt. Col. G.S.	July 1919	July 1919
Frink, J.L.	Lt. Col. G.S.	July 1919	July 1919
Peek, G.M.	Major, C.A.C.		
(Lt. Col. to 30 June 1920)		July 1919	----------
Dunn, W. McK	Capt. F.A.	July 1919	Aug. 1919
Moore, R.C.	Capt. F.A.	Aug. 1919	Dec. 1920
Kingman, A.F.	Capt. Inf.	Aug. 1919	----------
Ristine, H.H.	Capt. F.A.	Nov. 1919	Apr. 1920
Pinkerton, W.P.	Capt. Inf.	Jan. 1920	Nov. 1920
Attached.			
Batchelor, J.H.	Major, F.A. (Leave Off.)	July 1919	Sept 1919
Bennett, F.W.	Capt. USMC (Supt. Bldg)	July 1919	Sept 1919

Perkins, C.R.	Capt. Inf. (Supt. Bldg)	Sept 1919	May 1920
Gerow, L.S.	Capt. Inf. (Ath. & Wel. O)	July 1920	---------
Johnson, R.D.	Lt. Col. (Ath O)	July 1919	July 1919
Webzter, R.M.	Maj., Inf. (Vis Bureau)	July 1919	May 1920
Parr, L.C.	2nd Lt QMC (Vis Bureau)	July 1919	Feb. 1920
Bodecker, C.C.	Capt. Inf (Vis Bureau)	Sept 1919	Nov. 1920
Armstrong, F.T.	Major, F.A. (Vis Bureau)	Oct. 1919	Apr. 1921
Withers, A.P.	Capt. Inf. (Ath. Officer)	Sept 1919	Jan. 1921

A. C. OF S. G-2

Bagby, P. H. (Lt Col. to June 30, 1920)	Major, G.S.	July 1919	----------
Scully, A.M.	Major, Inf	July 1919	Aug. 1919
Arneman, G.E.	Major, F.A.	Sept 1919	Oct. 1920
Hossfeld, H.	Lt Col., Inf.	Aug. 1919	Dec. 1921
Berthold, H.A.	Capt., C.A.C.	July 1919	Dec. 1920
Dearden, W.H.	Capt., Inf	July 1919	Sept 1919
Nourse, E.M.	Capt., M.T.C.	July 1919	Sept 1919
Wernecke, O.F.	Capt., Inf.	July 1919	Aug. 1919
Lamm, G.H.	1st Lt. Inf.	July 1919	Apr. 1920
Stern, M.J.	1st Lt. Inf.	July 1919	Sept 1919
Davis, N.F.	2nd Lt. C. of I.	July 1919	Aug. 1919
Holzmann, A.W.	2nd Lt. C. of I.	July 1919	Nov. 1920
Beitz, W.E.	Capt., F.A.	Sept 1919	Apr. 1921
McMahon, B.B.	Capt., Inf.	Sept 1919	----------
Osann, H.E.	Capt., Inf.	July 1919	----------
Coulter, H.K.	Capt., Inf.	Aug. 1919	Jan. 1921
Minuth, H.C.	Capt., Cav.	Aug. 1919	July 1920
Switzer, H.C.	1st Lt., Inf.	Sept 1919	Feb. 1920
Savage, E.W.	Capt., Inf	Sept 1919	Oct. 1920
Bollenbeck, J.W.	Capt., Inf	Oct. 1919	Feb. 1921
Gardner, J.H. Jr.	1st Lt., F.A.	Aug. 1919	May 1921
Taber, T.R.	1st Lt., Cav.	Aug. 1919	Aug. 1920
Balcar, P.G.	1st Lt., Inf.	Aug. 1919	Dec. 1920
Teall, P.H.	2nd Lt., C. of I.	July 1919	July 1920
Bird, C.W.	1st Lt., C. of I.	July 1919	Dec. 1920
Gillis, M.A.	1st Lt., Inf.	Dec. 1920	Feb. 1921
Schoonmaker, L.E.	Capt., C.A.C.	Feb. 1921	----------
Dedicke, A.E.	Capt., Inf.	Dec. 1921	----------

Attached.

Rice, E.F.	Lt. Col., G.S.	July 1919	Aug. 1919
Seebach, O.	Lt. Col., Inf.	July 1919	Aug. 1919
Fernbach, O.H.	Major, C. of I.	July 1919	Sept 1919
Cordes, E.	Capt., Inf.	July 1919	Aug. 1919
Townsend, R.F.	Capt., Eng.	July 1919	July 1919
Stroup, A.R.	Capt., Eng.	July 1919	Aug. 1919
Berryhill, F.L.	Capt., Inf.	July 1919	Aug. 1919

Geedike, J.G.	Capt., Eng.	July 1919	Aug. 1919
Holt, J.R.	1st Lt. Inf.	July 1919	Aug. 1919
Falk, E.H.	1st Lt. F.A.	July 1919	July 1919
Charters, J.J.	1st Lt., Eng.	July 1919	Aug. 1919
Aldridge, A.H.	1st Lt., Inf.	July 1919	Aug. 1919
Church, H.W.	1st Lt., C. of I.	July 1919	July 1919
Distelhorst, F.M.	Capt., Inf.	July 1919	Jan. 1920
Berg, H.	1st Lt., Inf.	July 1919	Sept 1919
Dieckhoff, J.C.	1st Lt., T.C.	July 1919	Oct. 1919
Norris, J.S.	1st Lt. G.S.	July 1919	Oct. 1919
Dirkson, E.M.	2nd Lt., F.A.	July 1919	Aug. 1919
Haigh, A.C.	2nd Lt., Inf.	July 1919	Aug. 1919
Florin, A.	2nd Lt., C. of I.	July 1919	Mar. 1920
Brunaugh, R.C.	2nd Lt., F.A.	July 1919	Aug. 1919
Bichwit, L.	2nd Lt., C. of I.	July 1919	Aug. 1919
Amaroc News.			
Mueller, Paul J.	Major, Inf.	Sept 1920	---------
Damm, T.	1st Lt., Inf.	July 1919	Sept 1919
Stafford, G.C.	2nd Lt., F.A.	July 1919	Dec. 1920
A. C. OF S., G-3			
Montgomery, J.C. (Lt. Col. to June 30, 1920)	Major, G.S.	July 1919	Apr. 1920
Wainwright, J.M.	Lt. Col.	July 1919	Sept 1920
Collins, J.L.	Capt., Inf.	June 1920	July 1921
Robinson, D.A.	Major, Cav.	Aug. 1920	Sept 1920
Lonergan, T.C.	Major, G.S.	Oct. 1920	Dec. 1920
Fechet, d'A	Major, Inf.	Dec. 1920	Sept 1921
Haislip, W.H.	Major, Inf.	Dec. 1920	July 1921
Herr, J.K.	Major, G.S.	Jan. 1921	----------
Downer, J.W.	Major, F.A.	Jan. 1921	----------
Hodges, C.B.	Major, G.S.	June 1921	Dec. 1921
McCarthy, L.J.	Capt., Inf.	June 1921	----------
McCulloch, W.A.	Major, Inf.	Sept 1921	----------
E. & V. Training.			
Young, F.S.	Lt Col., G.S.	Mar. 1920	July 1921
Johnson, H.	Capt., A.G.D.	May 1920	Dec. 1920
Houghland, H.J.	Capt., A.S.	May 1920	Jan. 1921
Heyser, M.C.	Capt., F.A.	June 1920	Feb. 1921
Townsend, S.A.	Capt., Cav.	July 1920	Aug. 1921
Johnson, W.B.	Capt., Q.M.C.	Sept 1920	Apr. 1921
McKee, R.G.	1st Lt. Inf.	Nov. 1920	Mar. 1921
Dutton, D.L.	Capt., C.A.C.	Dec. 1920	Feb. 1921
Campbell, W.F.	1st Lt., Q.M.C.	Jan. 1921	Mar. 1921
Moss, W.H.	Major, Inf.	July 1921	---------
Marshall, L.W.	Capt., Inf.	July 1921	---------
West, D.R.	1st Lt., Q.M.C.	Oct. 1921	---------
Liaison Officers.			
Coleman, LeV	Colonel, C.A.C. (Mainz)	Nov. 1919	---------
Biddle, D.H.	Col. F.A. (Cologne)	Nov. 1919	Jan. 1921

Longstreet, J.	Lt Col., Cav. (Cologne)	Jan. 1921	---------
Bartlett, G.G.	Major, Inf. (Wiesbaden)	Feb. 1920	---------
Williamson, R.	Capt., Engr. (Aix-La-Chapelle)	Nov. 1919	Jan. 1921
Minuth, H.C.	Capt., Cav. (Mainz & Aix-La-Chapelle)	July 1920	---------
Russell, G.M.	Col., F.A. (Wiesbaden)	Nov. 1919	Mar. 1920

A. C. OF S., G-4

Chaffin, A.D.	Lt Col. G.S.	July 1919	July 1919
Somervell, B.	Major, Engrs.	July 1919	Sept 1920
Robinson, D.A.	Major, Cav.	Sept 1920	Dec. 1920
Herr, J.K.	Major, G.S.	Nov. 1920	Feb. 1921
Sheen, H.H.	Lt Col. Q.M.C.	Jan. 1921	---------
Case, C.Z.	Major, Cav.	July 1919	Aug. 1919
Miller, R.L.	Capt., Q.M.C.	Sept. 1919	Jan. 1921
McGuire, E.C.	Major, Cav. (Auditor)	July 1919	July 1921
Patterson, R.B.	Major, Cav.	July 1921	---------
Platt, J.C. Jr.	Capt., Inf.	Dec. 1921	---------
Stewart, H.P.	1st Lt. Cav. (Auditor)	Nov. 1919	Oct. 1920

G-4 Services

Supply Branch

Howard, J.H.	Lt Col., G.S.	July 1919	July 1919
Harris, A.R.	Major, G.S.	Aug. 1919	Mar. 1920
McMahon, W.C.	Capt., Inf.	July 1919	Nov. 1920
Kells, C.H.	Capt., Inf.	July 1919	Sept 1919
Bowker, F.E., Jr.	Capt., Q.M.C.	July 1919	July 1919
Dalton, J.N.	Capt., Inf. (Civ Relief Com)	Aug. 1919	Nov. 1920

Transportation.

Woodruff, E.N.	Major, Engrs.	July 1919	July 1919
Crawford, J.B.	Major, C.A.C	May 1920	July 1920
Boone, A.	Major, Cav.	July 1920	Oct. 1920
McCain, W.A.	Major, Q.M.C.	Oct. 1920	---------
Bomar, E.C.	Capt., C.A.C.	July 1919	May 1920
Sadler, W.H.	Capt., Q.M.C.	July 1919	May 1920
Troland, G.B.	Capt., Engrs.	July 1919	May 1920
Courtney, M.A.	Capt., T.C.	Sept 1919	Oct. 1919
Creed, R.L.	Capt., Cav.	Oct. 1919	July 1921
Daugherty, R.D.	Capt., Inf.	Nov. 1919	May 1920

ADJUTANT GENERAL's DEPARTMENT.

Haverfield, W.A.	Lt Col. A.G.D.	July 1919	Aug. 1919
Talbott, S.G. (Lt Col. to April 1920)	Major, A.G.D.	Aug. 1919	---------
Nulsen, C.K.	Major, A.G.D.	Nov. 1920	---------
Thiele, C.M.	Major, A.G.D.	Nov. 1919	---------
Carlson, E.F.	Capt., F.A.	July 1919	Sept 1919

White, C.W.	Capt., F.A.	Aug. 1919	Sept 1920
Wolfson, H.	Capt., A.G.D.	July 1919	Aug. 1919
Lauer, W.F.	Capt., Inf.	Aug. 1919	---------
Kingston, H.L.	1st Lt., Inf.	July 1919	Aug. 1919
Doty, K.	1st Lt., Inf.		
	(Cables & Codes)	Aug. 1919	Aug. 1921

INSPECTOR GENERAL'S DEPARTMENT.

Read, A.C.	Colonel, I.G.D.	July 1919	Jan. 1920
Peck, R.C.	Colonel, I.G.D.	July 1919	July 1919
Rice, C.H.	Lt Col., I.G.D.	July 1919	Sept 1919
Vestal, S.C.	Colonel, I.G.D.	July 1919	July 1919
Young, F.S.	Lt Col., G.S.	July 1919	Sept 1919
Magruder, L.B.	Major, I.G.D.	Aug. 1919	---------
(Lt Col. to Mar. 15, 1920)			
Van Schaick, L.J.	Colonel, I.G.D.	Nov. 1919	July 1920
Streater, W.	Major, I.G.D.	Jan. 1920	Nov. 1920
Hay, W.H.	Colonel, I.G.D.	Aug. 1920	May 1921
Parkinson, J.L.	Major, I.G.D.	Dec. 1920	---------

JUDGE ADVOCATE GENERAL'S DEPARTMENT

Rucker, K.	Lt Col., J.A.	July 1919	Aug. 1921
(Colonel to June 30, 1920)			
Barnwell, N.	Lt Col., J.A.	July 1919	Aug. 1919
Tears, C.C.	Lt Col., J.A.	July 1919	Oct. 1919
Todd, C.C.	Colonel, Inf.	June 1920	---------
Winship, B.	Colonel, J.A.		
	(DS Paris)	Sept 1921	---------
Grinstead, R.E.	Lt Col., Q.M.C.	June 1921	July 1921
Ireton, R.E.	Capt., Inf.	Aug. 1919	Nov. 1920
Bush, W.H.	Major, J.A.D.	Jan. 1920	July 1920
Morrisette, J.E.	Capt., J.A.D.	Oct. 1920	---------
Steward, E.M.S.	Capt., Inf.	Oct. 1921	---------

QUARTERMASTER CORPS

Williamson, G.M.	Col., Q.M.C.		
	(Chief Q.M.)	July 1919	Sept 1919
Ogborn, D.	Major, Q.M.C.	July 1919	Sept 1919
Morell, F.	Major, Q.M.C.	July 1919	Nov. 1919
Grove, D.	Major, Q.M.C.	July 1919	Aug. 1919
Stewart, F.P.	Capt., Q.M.C.	July 1919	Sept 1919
Buckbee, E.J.	Major, Q.M.C.	July 1919	Sept 1919
Rock, W.F.	Capt., Q.M.C.	July 1919	Sept 1919
Wiles, H.F.	Capt., Inf.	July 1919	Oct. 1919
Falk, A.G.	2nd Lt. Q.M.C.	July 1919	Aug. 1919
Drury, R.W.	Lt Col., Q.M.C.	July 1919	Dec. 1919
Sheen, H.H.	Lt Col., Q.M.C.		
(Col. to June 30, 1920)		Sept 1919	---------
Longstreet, J.	Lt Col., Cav.	Aug. 1919	Aug. 1919
Knight, O.F.	1st Lt., F.A.	Aug. 1919	Mar. 1920
Rogers, F.C.	Major, Q.M.C.	Aug. 1919	July 1921
Hahn, J.B.	Lt Col., Q.M.C.	Sept 1919	Oct. 1919
Grinstead, R.E.	Lt Col., Inf.	Nov. 1919	Nov. 1920
McFall, J.H.	2nd Lt., Q.M.C.	Oct. 1919	Nov. 1919
Gilbet, C.L.	Major, C.A.C.	Nov. 1919	---------

Reynolds, C.C.	Lt Col., F.A.	Dec. 1919	Sept 1920
Rich, A.T.	Capt., Inf.	Nov. 1919	Dec. 1919
Lowenberg, M.A.	Capt., Cav.	Nov. 1919	Jan. 1920
Ming, M.A.S.	Capt., F.A.	Dec. 1919	Jan. 1920
Montgomery, E.	Major, C.A.C.	Jan. 1920	June 1921
Young, J.E.	1st Lt. Inf.	April 1920	May 1920
Chapman, G.F.	1st Lt., Q.M.C.	June 1920	Nov. 1920
Sturges, E.A.	Lt Col., Fin. Dept.	Dec. 1919	Jan. 1920
Rosen, H.I.	1st Lt., Q.M.C.	Oct. 1920	---------
McGrath, T.J.	Capt., Q.M.C.	Oct. 1920	---------
Willford, J.W.	1st Lt., Q.M.C.	Oct. 1920	Dec. 1920
Hanna, C.W.	1st Lt., Inf.	Dec. 1920	Apr. 1921
McMahon, W.C.	Capt., Inf.	Nov. 1920	Aug. 1921
Bennett, E.E.	Major, C.A.C.	Jan. 1921	Mar. 1921
Scott, E.M.	Capt., Q.M.C.	June 1921	---------
Herwig, H.R.W.	Major, Q.M.C.	July 1921	---------
Archer, S.D.	1st Lt., Q.M.C.	Oct. 1921	Dec. 1921
Kidwell, G.B.	1st Lt., Q.M.C.	Oct. 1921	---------
Disbursing Q.M. (Changed to Finance Department, July 1920)			
Baker, F.J.	Major, Q.M.C. (Disb Q.M.)	July 1919	Oct. 1919
Zautner, H.G.	Major, Q.M.C.	July 1919	Aug. 1919
Lockard, E.E.	Major, Fin. Dept.	July 1919	---------
Walton, J.H.	Major, Q.M.C.	July 1919	June 1920
Green, F.E.	Capt., Q.M.C.	July 1919	Aug. 1919
Barksdale, J.A.	1st Lt., Q.M.C.	July 1919	Aug. 1919
Patton, F.	2nd Lt., Cav.	Sept 1919	Nov. 1919
McFall, J.H.	1st Lt., Finance Dept. (Auditor)	Nov. 1920	---------
Rice, H.G.	Major, Fin. Dept.	Nov. 1919	---------
Sturges, E.A.	Lt Col., Finance Dept. (Chief Fin.O)	Aug. 1920	---------
Enders, E.M.	Major, Fin. Dept. (Auditor)	June 1920	Dec. 1921
Stewart, H.P.	1st Lt., Cav. (Auditor)	Oct. 1920	Mar. 1921
MOTOR TRANSPORT SERVICE			
Graham, A.M.	Col. Cav. (Chief MTO)	July 1919	Nov. 1919
Christy, W.C.	Lt. Col., M.T.C.	July 1919	Nov. 1919
Couchman, A.W.	Major, M.T.C.	July 1919	Nov. 1919
Kasper, J.A.	Capt., M.T.C.	July 1919	Feb. 1920
Cowan, F.R.	2nd Lt., M.T.C.	July 1919	Sept 1919
Reed, J.M.	2nd Lt., M.T.C.	July 1919	Feb. 1920
DeBolt, E.E.	2nd Lt., M.T.C.	July 1919	Nov. 1919
Johnson, W.B.	Capt., Q.M.C.	Aug. 1919	Aug. 1920
Sheets, A.M.	Capt., T.C.	Aug. 1919	Mar. 1920
Martin, W.B.	1st Lt., M.T.C.	Aug. 1919	Jan. 1920
McIlwain, S.W.	Capt., M.T.C.	Aug. 1919	Dec. 1919
Lee, F.C.	2nd Lt., M.T.C.	Aug. 1919	Oct. 1919
Mahoney, C.V.	1st Lt., M.T.C.	Aug. 1919	Sept 1919

Roach, L.L.	Lt Col., Inf. (Chief MTO)	Nov. 1919	Sept 1920
Dalmore, L.	Capt., M.T.C.	Nov. 1919	Oct. 1920
Reeder, H.L.	Capt., Inf.	Apr. 1920	Dec. 1920
Weeks, H.J.	Major, Q.M.C. (Chief MTO)	Sept 1920	---------
Reed, F.F.	Capt., Q.M.C.	Oct. 1920	Apr. 1921
Newland, J.T.	1st Lt., Q.M.C.	Nov. 1920	Jan. 1921
Liquidation Comm. General Sales			
Graves, E.	Col., Engrs.	July 1919	Aug. 1919
Harvey, W. Jr.	Capt., A.S.	July 1919	Sept 1919
Coleman, LeV	Col., C.A.C.	July 1919	Sept 1919
Kruttschnitt, T.H.	Lt. Col., Q.M.C.	Sept 1919	June 1920
Downer, J.W.	Lt. Col., F.A.	Sept 1919	Dec. 1919
Wilson, B.L.	Major, M.C.	Nov. 1919	Jan. 1920
Callahan, J.W.	Capt., C.A.C.	Nov. 1919	Feb. 1920
Holland, E.R.	Capt., C.A.C.	Nov. 1919	Nov. 1919
Bolibaugh, O.B.	Capt., M.C.	Nov. 1919	Nov. 1919
Griebel, F.P.P.	1st Lt., San. C.	Nov. 1919	Feb. 1920
Smith, R.	Col., Inf.	Dec. 1919	Jan. 1920
Remount Service			
Flint, H.A. (Lt. Col. to March 15, 1920)	Major, F.A.	July 1919	Dec. 1920
Gardner, H.F.	Capt., Q.M.C.	July 1919	Dec. 1919
Gailey, J.K.	1st Lt., Cav.	Aug. 1919	Oct. 1919
Lynch, E.F.	2nd Lt., Q.M.C.	Sept 1919	Nov. 1919
Downer, J.W. (Lt. Col. to March 15, 1920)	Major, F.A.	Jan. 1921	---------
Salvage Service.			
Chambers, H.C.	Major, Q.M.C.	July 1919	Sept 1919
Albright, R.F.	1st Lt., Q.M.C.	July 1919	Sept 1919
Phelan, R.V.	2nd Lt., Q.M.C.	July 1919	July 1919
Shines, T.B.	Capt., Q.M.C.	July 1919	Aug. 1919
MEDICAL DEPARTMENT			
Grissinger, J.W.	Col. M.C. (Chief Surgeon)	July 1919	July 1919
Boak, S.D.	Col., D.C. (Chief Dent. Surg.)	July 1919	Aug. 1919
Borden, W.B.	Lt. Col., M.C. (Med. Supply O)	July 1919	Oct. 1919
Ferenbaugh, T.S.	Lt. Col., M.C.	July 1919	Aug. 1919
Nelson, A.	Capt., San. C	July 1919	Aug. 1919
Bruns, E.H.	Lt Col., M.C. (Chief Surgeon)	July 1919	Oct. 1920
Bunce, J.W.	Major, M.C.	Nov. 1919	Nov. 1919
Johnson, J.F.	Major, M.C.	Nov. 1919	Jan. 1921

McDiarmid, N.L.	Major, M.C. (Medical Supply O)	April 1920	June 1920
Keefer, F.R.	Col., N.C. (Sur.)	Sept 1920	---------
Dear, W.R.	Major, M.C.	Oct. 1920	Oct. 1921
Scott, H.O.	Lt., Col., D.C.	Dec. 1920	---------
Cleaves, J.W.	1st Lt., M.A.C.	Jan. 1921	---------

Attending Surgeon's Office.

Harrington, J.	Major, D.C.	July 1919	July 1919
Dyson, J.E.	1st Lt., M.C.	July 1919	July 1919
Steele, J.M.	Major, M.C.	July 1919	Aug. 1919
Skelton, O.G.	Major, D.C.	Aug. 1919	Dec. 1920
Felch, M.F.	Major, M.C.	Sept.1919	Dec. 1920
Newson, B.I.	Major, D.C.	Nov. 1919	July 1920
Dean, B.H.	Capt., D.C.	Nov. 1919	Oct. 1920
Thompson, R.F.	Capt., D.C.	Nov. 1919	Apr. 1921
Edwards, J.F.	Capt., M.C.	Nov. 1919	Jan. 1921
Hartley, L.P.	Capt., D.C.	Nov. 1919	Jan. 1921
Gray, D.S.	Capt., D.C.	Nov. 1919	Oct. 1920
Weaver, M.S.	Capt., M.C.	Nov. 1919	Feb. 1921
Rose, W.A.	Capt., D.C.	Dec. 1919	Apr. 1921
Parsons, W.B.	1st Lt., D.C.	July 1920	Oct. 1920
Rohde, S.J.	Capt., D.C.	Sept 1920	Mar. 1921
Tyng, F.C.	Capt., M.C.	Feb. 1921	Apr. 1921
Price, H.H.	Capt., M.C.	Jan. 1921	Apr. 1921

Veterinary Corps.

Gould, J.H.	Lt Col., V.C. (Chief Vet.)	July 1919	Aug. 1919
Renshaw, S.B.	Capt., V.C. (Chief Vet.)	Aug. 1919	Dec. 1920
Derrick, G. W.	1st Lt., V.C.	Oct. 1919	May 1920
Miner, J.W.	1st Lt., V.C.	Nov. 1919	Jan. 1921
Feazel, G.F.	1st Lt., V.C.	Nov. 1919	Dec. 1919
Casey, J.W.	1st Lt., V.C.	Nov. 1919	Dec. 1919
Shook, L.L.	1st Lt., V.C.	Jan. 1920	Mar. 1920
Foster, R.J.	Major, V.C.	Nov. 1920	---------
Austin, F.M.	1st Lt., V.C.	Dec. 1920	Apr. 1921

CORPS OF ENGINEERS.

Stewart, J.W.	Lt Col., Engrs. (Chief Engr.)	July 1919	Sept 1919
Morris, V.	Major, Engrs.	July 1919	July 1919
Whitney, C.E.	Major, Engrs.	July 1919	Oct. 1919
Post, W.L.	Major, Engrs.	July 1919	Aug. 1919
Lynde, C.C.	Capt., Engrs.	July 1919	Aug. 1919
Strang, J.A.	Capt. Engrs.	July 1919	Oct. 1920
Bowler, W.E.	1st Lt., Engrs.	July 1919	Sept 1919
Maxon, W.R.	2nd Lt., Engrs.	July 1919	Sept.1920
Keesler, A.A.	Capt., Engrs.	July 1919	Aug. 1919
Miller, B.A.	Major, Engrs (Lt Col. to June 30, 1920)	Sept 1919	Dec. 1921
Lyons, F.R.	Capt., Engrs.	Sept 1919	Oct. 1920
Frech, F.F.	1st Lt., Engrs.	Sept 1919	Oct. 1919
Marschner, W.B.	2nd Lt., Engrs.	Aug. 1919	Oct. 1919

Drinkwater, J.G.	Capt., Engrs.	Sept 1920	---------
Munroe, R.K.	Capt., Engrs.	Sept 1920	---------
Joly, C.L.	Major, C.W.S.	May 1921	---------
Bethel, E.A.	Major, Engrs.	Dec. 1921	---------

ORDNANCE DEPARTMENT

Crain, J.K. (Lt Col. to June 30, 1920)	Major, Ord. Capt.,	July 1919	July 1921
Schimelfenig, C.A. (Lt Col. to June 30, 1920)	Major, Ord. Dept.	July 1919	Aug. 1919
Gage, P.S.	Lt Col., Cav.	July 1919	July 1919
McIntyre, H.C.	Major, Ord, Dept.	July 1919	July 1919
Handy, M.	Capt., Ord. Dept.	July 1919	Aug. 1919
Hofstetter, C.F.	1st Lt., Ord. Dept.	Oct. 1920	Nov. 1920
Maxwell, R.L.	Major, Ord. Dept.	June 1921	July 1921
Jordan, H.B.	Lt Col., Ord. Dept.	July 1921	---------

SIGNAL CORPS.

Hemphill, J.E.	Col. S.C. (Chief Sig.O)	July 1919	Aug. 1919
Gauss, W.F.	Major, S.C.	July 1919	Sept 1919
Temple, H.H.	Capt., S.C.	July 1919	Oct. 1920
Romesburg, L.	Capt., S.C.	July 1919	Dec. 1920
Kreis, R.	Capt., S.C.	July 1919	Oct. 1919
Drake, C.	Capt., Inf.	July 1919	July 1919
Granger, R.M.	1st Lt., S.C.	July 1919	Dec. 1920
Sailor, V.L.	1st Lt., Inf.	July 1919	July 1919
Evans, W.	2nd Lt., S.C.	July 1919	July 1919
Burton, C.V.	1st Lt., S.C.	July 1919	Sept 1919
Ware, J.F. (Lt. Col. to June 30, 1920)	Major, S.C.	Aug. 1919	---------
Dolan, J.A.	1st Lt. S.C.	Aug. 1919	Oct. 1919
Wells, E.R.	1st Lt., S.C.	Oct. 1919	---------
Merkel, W.C.	Capt., S.C.	April 1920	---------

AIR SERVICE

Fowler, H.	Col., A.S.	July 1919	July 1919
Harmon, H.R.	Major, A.S.	July 1919	Sept 1919
Berry, A.W.	Major, A.S.	July 1919	July 1919
Andrews, F.M.	Major, A.S.	Sept 1920	---------
Murphy, W.H.	Capt., A.S.	Oct. 1920	Jan. 1921
Jervey, T.M.	1st Lt., Inf.	Oct. 1920	Nov. 1920
Guenther, K.D.	1st Lt., A.S.	Jan. 1921	Mar. 1921

CHAPLAINS

Ferguson, L.R.S.	Chap. (Sr. Chap.)	July 1919	July 1919
Easterbrook, E.P.	Lt. Col., Chap (Sr. Chap.)	Mar. 1921	---------

R. R. & CLAIMS SERVICE

Johnson, A.A.	Capt. Inf.	July 1919	Aug. 1919
Donough, R.P.	1st Lt. Engrs.	July 1919	Sept 1919
Foster, F.A.	2nd Lt. F.A.	July 1919	Aug. 1919
Wogan, J.B. (Major to June 30, 1920)	Capt., C.A.C.	Aug. 1919	Oct. 1920
Mahon, S.	Capt., F.S.	Aug. 1919	Feb. 1921

Meske, F.	Capt., Inf.	Aug. 1919	Oct. 1919
Coyne, F.E.	2nd Lt., F.A.	Aug. 1919	Sept 1919
Wolber, C.K.	Capt., Inf.	Oct. 1919	---------
Tucker, J. D.	1st Lt., Inf.	Oct. 1919	Nov. 1919
Bennett, J.	2nd Lt., Inf.	Oct. 1919	Nov. 1919
Bink, J.A.	2nd Lt., Inf.	Oct. 1919	Nov. 1919
Galloway, J.W.	2nd Lt., Inf.	Oct. 1919	Nov. 1919
Herrick, D.B.	2nd Lt., Inf.	Oct. 1919	Nov. 1919
Shanks, C.L.	2nd Lt., Inf.	Oct. 1919	Nov. 1919
Ware, J.V.	Major, Inf.	Dec. 1919	Dec. 1919
Teall, H.P.	2nd Lt., C. of I.	Jan. 1920	Apr. 1920
Skene, C.M.S.	Major, C.A.C.	Oct. 1920	Dec. 1920
Donaldson, W.H.	Capt., C.A.C.	Oct. 1920	---------
Hanna, L.M.	1st Lt., C.A.C.	Dec. 1920	---------
Bodeker, C.C.	Capt., Inf.	Feb. 1921	---------
Ming, M.A.S.	Capt., F.A.	Mar. 1921	---------

PROVOST MARSHAL'S DEPARTMENT

Cockriel, G. (Major to June 30, 1920)	Capt., M.P.C.	July 1919	Dec. 1920
Jenne, F.H.	Capt., Inf.	July 1919	July 1919
Flint, A.M.	Capt., Inf.	July 1919	July 1919
Lauritsen, P.	1st Lt., Inf.	Aug. 1919	Dec. 1920
Henderson, D.L.	Major, Inf.	Aug. 1919	Nov. 1919
Manner, M.M.	Capt. Inf.	Nov. 1919	Nov. 1919
Dewhurst, W.	Capt., Cav.	Nov. 1919	Nov. 1919
Moore, G.	1st Lt., Inf.	Nov. 1919	Nov. 1919
McKittrick, W.K.	1st Lt., Inf.	Nov. 1919	Nov. 1919
Munday, B.F.	Capt., Cav.	Nov. 1919	Nov. 1919

Note - This department was transferred to Hq. Commandant on December 21, 1920.

CIVIL AFFAIRS.

Hunt, I.L.	Col., Inf.	July 1919	Mar. 1920
Gilbert, W.O.	Col., J.A.	July 1919	Oct. 1919
Bensel, W.	Lt. Col., M.C.	July 1919	Jan. 1920
Evans, V.	Major., Inf.	July 1919	Mar. 1920
Smith, T.	Major, Inf.	July 1919	June 1920
McKey, J.R.	Capt., Inf.	July 1919	Sept 1919
Chapman, M.D.	Capt. U.S.M.C.	July 1919	Aug. 1919
Parker, C.	Capt., Inf.	July 1919	July 1919
Barker, T.H.	Capt., Inf.	July 1919	Jan. 1920
Tower, W.L.	Capt., San. C.	Aug. 1919	Feb. 1920
Lockwood, B.C.	Major, Inf.	Aug. 1919	Sept 1919
White, J.D.	Capt., F.A.	Sept 1919	Dec. 1919
Davis, M.	Major, Inf.	Aug. 1919	Oct. 1919
Hall, R.P. (Lt Col. to March 15, 1920)	Major, C.A.C.	Oct. 1919	Sept 1920
Pinkerton, H.M.	Capt., Inf.	Oct. 1919	Feb. 1920
Harper, P.C.	Capt., F.A.	Nov. 1919	Feb. 1921
Melberg, R.	Major, C.A.C.	Dec. 1919	Oct. 1920
Stayer, M.C. (Col. to Feb. 17, 1920)	Major, M.C.	Jan. 1920	Sept 1920
Goff, G.D.	Col., J.A.D.	Feb. 1920	Apr. 1920
Grier, H.S.	Lt Col., G.S.	Mar. 1920	July 1920

Bente, C.F.	1st Lt., Inf.	Mar. 1920	June 1920
Bird, C.W.	1st Lt., C. of I.	Oct. 1919	June 1920
Dunlop, R.H.	Major, Inf.	Nov. 1919	Aug. 1920
Hackert, C.H.	2nd Lt., C. of I	Sept 1919	Oct. 1920
Lammons, F.B.	Capt., Inf.	Aug. 1919	Sept.1920
Flynn, T.J.	Major, M.C.	Sept 1920	---------
Barker, J.W.	Capt., C.A.C.	Oct. 1920	Nov. 1921
Bevans, S.M.	1st Lt. F.A.	Dec. 1920	Apr. 1921
Goodrich, P.M.	Lt Col., Inf.	Mar. 1921	---------
Downs, S.D. Jr.	Major, F.A.	May 1921	---------
Crea, H.B.	Major, Inf.	Nov. 1921	---------
Stone, D.L.	Colonel, G.S.	Sept 1920	---------
(Military Advisor Inter-Allied R.H. Com. since January 25, 1920)			

HEADQUARTERS COMMANDANT.

McCammon, W.W.,	Lt. Col., Inf.	Dec. 1920	---------
Helsley, A.B.	Capt., Inf.	Dec. 1920	---------

CHAPTER NO. 17

DISCIPLINE AND MORALE.

General Remarks; Discipline and Morale in the Third Army - the "Get Home: Desire; Supervision - Agencies; Conditions in the A.F.G.; Report of Commanding General, June 30, 1920; Provost Marshal's Department - Disciplinary Barracks, Military Prison, Division of Criminal Investigation; Senior Chaplain; Welfare Organizations; Officers.

DISCIPLINE AND MORALE.

The successful administration of an army of occupation requires the solution of manifold problems. Not the simplest of these is the establishment and maintenance throughout the command of a high standard of discipline and of its concomitant, a high morale. No military force can achieve true efficiency unless its members are well disciplined and have a high morale. Good leadership is unhesitating in the use of measures necessary to the institution of that discipline which is indispensable to military efficiency. It is not less active in measures for the comfort and well being of the individual. Discipline and the soldierly pride that is its consequent, contentment, and confidence in the leader - the inculcation of these in the soldier is essential to the establishment of a high morale in a military organization. Good discipline and high morale are complementary and one cannot exist in the absence of the other.

The necessity for good discipline and high morale is greater in an army of occupation than in other commands. The laws of modern war and the spirit of modern civilization alike prohibit the plunder and pillage of an invaded country. In the twentieth century the customs of war require that the inhabitants of an occupied territory be treated with all the consideration that the attendant circumstances will permit. In the American Army of Occupation those responsible for its discipline and morale were ever mindful of the fact that largely by the conduct of its soldiers would Europe appraise the American people. The American soldier in the Rhineland was constantly reminded by his leaders that upon him, his conduct, his attitude towards the conquered foe, and upon his dicipline and morale, rested in large part the good name not only of the American Army of Occupation but also of the United States Army and of his country. In addition, it was felt by the American Headquarters that as long as American troops remained on the Rhine, their efficiency as compared with European armies must be worthy of the great nation whose representatives they were.

The Third American Army arrived in its section of occupied Germany in December 1918. It was composed of combat divisions, all of which had been but recently engaged in the fierce fighting in the Argonne. Most of the soldiers had gone through the great excitement and strain that only battle affords. For weeks at a time during the recent past they had lived in dug-outs or in "fox-holes," sleeping in the mud and the muck, with lice and rats for their constant companions. For months they had been denied most of the ordinary comforts of modern life, and often they had been face to face with death, sometimes for days on end. They had witnessed violence and death in innumerable forms, and they had seen good comrades go to their Maker in the twinkling of an eye. They had been exhausted - physically, mentally and emotionally. With little chance for rest or recreation and at the conclusion of weeks, in some cases months, of great strain they had started on the march to the Rhine. Necessarily under close restrictions while en route, the American soldiers took up their stations in the Rhineland with a very human need and desire for relaxation. Under such conditions men are very apt to resort to vice, sexual as well as alcoholic. There was, then, an immediate and imperative need for the provision of facilities for wholesome amusement and recreation.

The maintenance of discipline, although not the iron discipline of the battlefield, was no less necessary. The soldier was separated by thousands of miles from the supporting and restraining influence of his home and his own country. In a strange land, among a people whose ideals and customs were different from his own and whose language he could not speak, he was forced to find his own amusement unless it was provided for him. Under such conditions men are all too prone to seek the recreation that is almost always available - liquor, loose women and gambling. Due to the depreciated state of the German currency, these vices were made doubly accessible to the soldier. It is an unfortunate fact that when large numbers of men of all classes are grouped together under conditions of enforced intimacy, there is a strong tendency towards a levelling down rather than up - in a barracks the coarser element is very apt to create the atmosphere of the group. The tendency mentioned is increased by loneliness, idleness and lack of recreation. Hence, in addition to the provisions of facilities for wholesome recreation, it was necessary that a high standard of discipline be maintained and that the soldier be kept busy.[1]

There was another factor that in many cases had a strong tendency to depress the spirits of the soldier and to increase his restlessness. This was a form of homesickness that took the shape of an unreasoning desire for an immediate return to the United States. Where this feeling was present in the mind of the soldier, he recked not of the military and political situation or of the limits of transportation facilities. He did not pause to consider that there were over two million American soldiers in Europe and that,

[1] Volume II, Chapter 19.

even under the pressure of imperative military necessity, an average of ten thousand a day was about the greatest number it had been possible to transport there. He wanted to "get home," his desire was insistent and impelling - and to him nothing else mattered. An American newspaper (published in Paris), apparently without heed to any other consideration than sensationalism and the increase of its circulation, instituted a "Get the Boys Home" campaign. In every issue the newspaper carried those words across the upper portion of its front page, while its columns contained articles in elaboration of the theme. It is difficult at best to maintain good discipline and morale in a great military force which is subjected to the let-down that is inevitable during an armistice. It is particularly difficult when the attendant circumstances are similar to those under which the American Expeditionary Forces labored after the conclusion of the Armistice Agreement of November 11, 1918. It is hardly too much to say that the ill-considered action of this American newspaper was the strongest of all factors in the causing of the lowered morale that was so perceptible in the A.E.F. in the winter of 1918 - 19.[1] It is certain that the "Get the Boys Home" campaign added greatly to the difficulties of the American officers in their efforts to maintain good discipline and morale. This insistent desire for an immediate return to the United States - it affected many of the younger officers too - was a far more potent factor in the morale problem among the troops in France than it was among those in Germany. The Third American Army was in occupation

[1] Upon informal inquiry as to the reason for the "Get the Boys Home" campaign, the editor replied that he had inaugurated it in order to "raise the morale" of the American soldiers.

of enemy territory; of itself this fact had great influence in retarding the development of the feeling under discussion. Again, that force was composed of most of the best combat divisions in the A.E.F. Nevertheless, the "get home" desire was present in the minds of many soldiers in the American area; it had to be reckoned with, and because of it the provision of wholesome recreation and amusement became all the more necessary.

The First Division of the General Staff (G-1) was charged with the supervision of the discipline and morale of the command and of all activities concerning those matters. It was specifically responsible for the provision of wholesome recreation and amusement for the troops. In the performance of these duties G-1 supervised and co-ordinated the activities of the following services and welfare organizations:

> Provost Marshal's Department, including Department of Criminal Investigation, Military Prison, and Disciplinary Barracks.
> Senior Chaplain,
> Young Men's Christian Association,
> Salvation Army,
> American Red Cross,
> Knights of Columbus,
> Jewish Welfare Board,
> American Library Association,
> Young Women's Christian Association.

The welfare organizations were semi-militarized. The personnel wore uniform (a special one for each organization), their movements were controlled by the military authorities, and they were subject to military orders and regulations.

After the disbandment of the Third Army, the Knights of Columbus and the Jewish Welfare Board withdrew their personnel from the American area, leaving five welfare organizations with the

American Forces in Germany. This chapter deals mainly with the discipline and morale of the A.F.G., but it may be said that during the existence of the Third Army the measures taken were effective and that good discipline and a high morale were maintained.

The American Forces in Germany, the command which succeeded the Third Army, came into existence on July 3, 1919. The problem relative to discipline and morale that confronted the A.F.G. and the means by which they were solved can best be illustrated by the following extracts from the annual report of the Commanding General, rendered to the War Department on June 30, 1920.

X X X X X

3. Personnel.

Upon my assuming command of these Forces, the commissioned and enlisted personnel consisted almost entirely of those who had served in the Third Army and so, in general, had participated in active operations as members of corps or divisions of the other armies of the A.E.F. The arrival of the 8th Infantry and of the Provisional Infantry Brigade, both of which units contained a very large proportion of recruits and of young officers without combat experience, and the gradual transfer to the United States of units and unattached personnel formerly belonging to the Third Army, created a situation wherein earnest and persevering endeavor upon the part of all experienced officers was necessary in order that there might emerge an efficient command of high discipline and morale. In my opinion the results achieved reflect credit upon the present commissioned and enlisted personnel of these Forces.

For a number of months there has been a gradual reduction in the strength of these Forces. Officers who, by their conduct, showed a lack of appreciation of their positions have been ordered to the United States, while many enlisted men have been similarly returned as undesirable. War Department orders have caused other officers to be relieved from duty here, while court-martial sentences and the normal death rate have still further reduced the enlisted strength.

X X X X X

9. Discipline and Morale.

The discipline of this command is under the circumstances very good, and while the degree fluctuated from time to time, the general trend has been steadily upward. During the first few months of the period covered by this report the securing of good discipline throughout the command constituted a difficult problem. Many of the men who had transferred from the combat divisions to units of the permanent garrison were undesirable characters, while the incoming units contained an unusually large proportion of recruits. In addition, a number of absentees and deserters from combat divisions remained in the small villages and rural districts of the American zone; only by a thorough combing process were they discovered and brought under military control. The facility and cheapness with which intoxicating drinks and immoral women could be secured furnished constant temptation to the soldier, and it was only through the earnest endeavors of officers, non-commissioned officers and the military police that a relatively high state of discipline was finally attained. Outwardly, in the matter of neatness, saluting and soldierly appearance in general, the soldiers of these Forces individually or in ranks are excellent and possibly are superior to those of other occupying forces. But in interior discipline, particularly as regards absenteeism, "very good" is the best rating that can be given the command as a whole. The establishment of a disciplinary barracks and the thorough course of training given the inmates have produced gratifying results, and have had a large part in raising the general disciplinary standard. Justice requires that mention be here made of the number of excellent soldiers, of every grade, included in this command. Their efficiency, bearing and exemplary conduct have given a tone to the whole command that has called forth the admiring comment of every competent observer.

The morale of these Forces is high, and there is a healthy spirit of rivalry and emulation in athletics as well as in drills, maneuvers and target practice between and among the various units. Frequent military and athletic competitions, some of them with associated forces, and the offering of prizes by welfare organizations to units having the best disciplinary records, all serve to keep alive this spirit and to help maintain the prevailing high morale.

X X X X X

11. Welfare Organizations.

X X X X Through the maintenance of clubrooms, reading and writing rooms, bowling alleys, restaurants, the furnishing of theatrical performances and moving pictures, and through the provision of material for athletic sports, these organizations have added greatly to the contentment of the command and have done much to maintain its morale.

X X X X X

> I desire to express my appreciation of their spirit of service and of the beneficial work performed by all these welfare associations. It is a fact that their efforts have done much to make more pleasant, to officers as well as to soldiers, the tour of service in occupied Germany.

X X X X X

During 1920 and 1921 the discipline and morale of the A.F.G., while occasionally displaying slight fluctuations, continued to be high. But their maintenance at that standard required constant care and well considered action upon the part of the responsible officers, supplemented by the unceasing activities of the various welfare organizations. Except during active operations, absenteeism and over-indulgence in strong drink have ever been the chief faults of the American soldier. The history of the A.E.F. shows that this was no less true of the volunteer and drafted man than, before and after the World War, it was of the peace time regular; perhaps, because of their failure (as shown by conditions in France during the months following the signing of the Armistice) to reach the Regular Army standard of discipline, the volunteers and drafted men as a whole were more prone to those faults than those Americans who, in time of peace, have composed the enlisted personnel of the Regular Army. However that may be, absenteeism and over-indulgence in strong drink were the main lapses from the required standard of discipline which the A.F.G. had to combat. Despite these faults and despite other difficulties inherent in any army of occupation, the A.F.G. were kept at a standard of discipline and training that was highly complimented by all qualified beholders. This must ever be a matter of pride to all who served therein, as it should be to the American people.

Before passing to a description of the various activities concerning the subject matter of this chapter, it may be said that the larger part of the A.F.G. was composed of excellent non-commissioned officers and men. But for their co-operation and soldierly example a high standard of discipline and morale could not have been attained.

PROVOST MARSHAL'S DEPARTMENT.

After the departure in July 1919 of the military police units of the Third Army, police and traffic control duties in the American zone were performed by the 139th and 141st Military Police Battalions. The strength of the Military Police was decreased gradually from August 1, 1919, until finally the entire American area was policed by the 278th Military Police Company. The combined Headquarters Troop and Military Police Detachment was formed on August 21, 1919. The Military Police Detachment consisted of 2 officers and 50 picked men of the 278th Military Police Company. This number soon proved to be entirely inadequate for the proper police of the area and was promptly increased to 5 officers and 200 enlisted men.

The Provisional Infantry Brigade arrived in the American area on November 5, 1919. This brigade had a large proportion of untrained and undisciplined recruits, and a further increase in the number of military police became necessary. The 246th Military Police Company, consisting of 1 officer and 75 enlisted men, was brought from Brest to Andernach for police duty in that section. This company lost its identity as such and was assigned to the Military

Police Company, A.F.G., its personnel being placed on duty with the Second Brigade, A.F.G., as the "Silesian" Brigade had been designated.

In order to decrease the number of military police on duty in the A.F.G., it was decided to turn over to the German police some of the duties performed by the military police. Accordingly, beginning in the latter part of 1919, routine traffic posts were assigned to the German police. These posts were handled in a satisfactory manner, and by the end of the year 1921 all traffic posts had been taken over by the German police. The Provost Marshal's Department exercised supervision over the German police and on special occasions, such as Inter-Allied horse show days, track and field meets, etc., assisted them by taking over the traffic control of the direct routes to athletic fields. The German police were also of material aid to the Division of Criminal Investigation, particularly in the handling of vice work. In May 1921 two members of the German police were detailed for vice work duty with the D.C.I.; their duties were performed in a very satisfactory manner. In general the co-operation of the German police was all that could be desired.

During the latter part of 1919 the Provost Marshal's Department was very active. Trials by courts-martial were numerous, the offenses being varied in character. A number of homicides were committed, the victims including German inhabitants as well as American soldiers; a few of these accused were charged with murder, but in most cases the charges alleged manslaughter, as the degree or state of intoxication of the accused did not warrant conviction of the greater crime. Robberies were numerous; they were committed mainly by men absent without leave, and usually the offenders had belonged to divisions which had returned to the United States. Larceny was the cause of numerous trials, several of which were for the theft of government property. Most of the crimes in the calendar were committed by American soldiers or ex-soldiers in occupied Germany in 1919. But strenuous efforts by the P.M.D. and organization commanders soon produced a noticeable improvement. It frequently happens that enlisted men who may be good soldiers in the United States are not suited for duty overseas, and it was found that there were some men of this character in the A.F.G. A system was inaugurated whereby such men could be returned to the United States for reassignment; the procedure as laid down in paragraph 148½, Army Regulations, was followed, and enlisted men who were not suited for service in Germany were returned to their own country. The application of this system and of vigorous disciplinary methods produced the desired results, and by the summer of 1920 the discipline of the command was very good and its morale excellent.

During the succeeding eighteen months the P.M.D. performed its ordinary duties and was of much assistance in the maintenance of discipline. Its duties included the following:

Maintenance of order.
Enforcement of regulations and orders.
Apprehension of deserters and absentees.
Investigation of crime and apprehension of criminals.
Liaison and co-operation with Allied military police.
Control and apprehension of prostitutes, vagrants and other undesirables.
Prevention of sale of prohibited intoxicants.
Recovery of stolen Government property.
Investigation of applications of soldiers to marry German or other foreign women.
Investigation of charges alleging abuse to American soldiers by individuals of the civil population.

Issue of passes to leave the American zone.

Control of circulation of military personnel.

Placing dubious resorts "Off Limits."

Liaison and co-operation with Military Intelligence Division, A.F.G.

Much assistance was rendered to the P.M.D. by the office of the Judge Advocate, which constantly impressed organization commanders with the necessity of resorting to the 104th Article of War in their administration of discipline; in other words, whenever practicable to administer "company punishment" rather than to prefer charges. The commander of each company unit was personally interviewed and given thorough instruction in the application of the 104th Article of War. Many offenses of a character that formerly had resulted in trial by Summary Court-Martial were disposed of by a wise imposition of summary punishment in accordance with the terms of this Article, and many first offenders and inexperienced soldiers were saved from the stigma of a trial by court-martial. Some young officers displayed a tendency to impose summary punishment for serious offenses that merited trial, but the general results were very good.

In order that men charged with offenses and awaiting trial should not be held in confinement for an unduly long period, frequent visits to all the guard houses in the area were made, and records and reports, especially charge sheets, were often examined; irregularities in charges and methods were corrected and all complaints were thoroughly investigated.

DISCIPLINARY BARRACKS.

After the departure of the Third Army units it soon became apparent that there was need in the A.F.G. for an institution similar to the disciplinary barracks in the United States. The Provisional Disciplinary Barracks, for the detention of certain military misdemeanants, was established on November 13, 1919. It was a potent factor in the maintenance of discipline in the A.F.G. The average monthly strength of those detained was about 120. The policy as to the organization of this institution and the classes of individuals to be there detained, permitted the admission of men sentenced to confinement at hard labor for not less than one month and not more than six months. Where a suspension of such sentence had been directed and under the suspended sentence there had been misconduct, the order of suspension was revoked and the offender was sent to the Disciplinary Barracks for the execution of the sentence. In addition to these offendors, organization commanders were permitted to send there men with sentences of not less than one month and not more than six months confinement, who, while serving sentence with their organization, had committed other offences and had shown by conduct that their association with selfrespecting soldiers was not in the interests of the service.

Soldiers in the Disciplinary Barracks were put through a thorough course of drill and instruction. A progressive training schedule was carried out under the supervision of the officer in charge of training. Due to the constant changes among the inmates, the program of training was not so extensive as that of an infantry company. Drill extended through the school of the soldier to the school of the squad as prescribed in the Infantry Drill Regulations, and great care was exercised as to the exactness in details. After approximately two months, should the soldier's conduct, application to duty, and knowledge of Infantry Drill Regulations, Manual of

Interior Guard Duty, and of the care of his equipment warrant, he was placed in a model platoon. If he then gave evidence of further improvement he was recommended for return to his organization under paragraph 943, Army Regulations. The system in force proved very effective and the Disciplinary Barracks continued in existence throughout 1921. Because of the great reduction in the strength of the A.F.G. scheduled for the early months of 1922, it was planned to abandon the Disciplinary Barracks in January of that year.

MILITARY PRISON.

Prior to November 1919, the Military Prison was under the administration of the Casual Depot, A.F.G. On November 10, 1919, the Military Prison Detachment was formed from men transferred from the Casual Depot, the Detachment consisting of 2 officers and 65 enlisted men. The prison was located in the old German Military Prison building on Fischel Strasse, Coblenz. Exclusive of quarters for the guards, it had a capacity of 300 beds. Only those soldiers charged with very serious offenses were confined in the Military Prison. The great majority of its inmates were Germans, and an average of 100 of these prisoners were sent daily to various organizations for fatigue work.

DIVISION OF CRIMINAL INVESTIGATION.

The Division of Criminal Investigation (Provost Marshal's Department) was organized in August 1919. Its original personnel consisted of 1 officer and 6 men who had been transferred from Base Section No. 9 at Antwerp.

The duties of the Division of Criminal Investigation (generally known as the D.C.I.) included the investigation of the following:

> Crimes committed by American soldiers and by civilians subject to the Articles of War.
>
> Crimes against the A.F.G. or members of the Allied community, committed by Germans of other nationals.

The D.C.I. co-operated with the Military Intelligence Division of the A.F.G. in bringing to trail those Germans or other nationals who had violated Ordinances of the Inter-Allied Rhineland High Commission or military regulations relative to the civil population. Its operatives, in conjunction with the German police, frequently arrested American or other offenders who had sought refuge in unoccupied Germany, and in many cases they recovered Government property which had been taken there. The lost or stolen Government and personal property recovered by the D.C.I. amounted to several millions of marks in value.

SENIOR CHAPLAIN.

The Senior Chaplain was charged with the supervision and co-ordination of the activities of all chaplains in the A.F.G. In addition to their religious and moral duties, the chaplains took an active part in the promotion of athletics and educational work in their respective organizations. Their relations with the welfare organizations were harmonious, and the mutual co-operation produced very satisfactory results. The chaplains were instructed that the general welfare of the soldiers was their chief concern, and regular visits to hospitals and military prisons were made.

One of the most important phases of the work of the Senior Chaplain related to the marriage problem.[1] Chaplains were forbidden

1 Volume II, Chapter 4.

to marry soldiers who could not produce a certificate showing that the German civil ceremony had been performed. Usually the soldier requested the chaplain of his organization to conduct the religious ceremony, but many asked for the services of the Senior Chaplain. Where this was not done, in the cases of those who reported to the Senior Chaplain for marriage, the latter assigned some other chaplain, Protestant or Roman Catholic as the case might be, to perform the ceremony. Records of all marriage ceremonies were preserved.

Numbers of soldiers, whose applications for permission to marry had been disapproved, brought their cases to the Senior Chaplain. If investigation indicated it to be advisable the latter requested that the case be reconsidered. Where this was done, the decision in most cases was changed and the soldier was permitted to marry. A request for reconsideration was never made unless the character of the woman was found by the Senior Chaplain to be relatively good. On the other hand, there were a few cases in which he intervened to prevent the marriage of an undesirable soldier to a woman of good character.

There were occasional conferences, upon questions of mutual interest, of the American chaplains with those of the Belgian, British and French armies of occupation. These conferences took place at Coblens or Cologne and were attended by chaplains of the Roman Catholic faith and of all Protestant denominations.

WELFARE ORGANIZATIONS.

The influence of the welfare organizations upon the discipline and morale of the A.F.G. could hardly be over-estimated. Primarily their mission was to furnish the soldier during his hours of idle-ness with wholesome amusement and recreation. No attempt was made to restrict vice, as this was a function of the Army. The endeavor was to provide such attractions in the way of entertainments, athletics, sports, etc., as to cause vice to be pushed into the background. That the activities of the welfare organizations largely fulfilled their purpose was shown by the great proportion of soldiers who took advantage of the opportunities afforded them.

In order that there should be no duplication of effort, the welfare organizations were placed on a military basis. Their activities were coordinated by the Welfare Officer (in G-1), who prescribed the points at which service should be given, established policies and furnished such support and assistance as was necessary and practicable. This system proved highly successful. Cordial relation sprevailed among the various societies and their co-operation with the military authorities produced excellent results. The ac5ivities of the welfare organizations had a large part in the production of the remarkable spirit of contentment that prevailed among the soldiers of the A.F.G. Their good work was often commended, and it met with the grateful appreciation of all concerned.

YOUNG MEN'S CHRISTIAN ASSOCIATION.

The vanguard of the "Y" forces came into the American zone on Decemver 11. 1918. By April 1, 1919, the "Y" had in operation 425 full-time regulation sercices, and 83 other points were served at frequent intervals by portable motion picture equipment and rolling canteens.

Theatrical entertainment seemed the form of service in the greatest demand, and the first units of professional entertainers arrived on Decemver 17, 1918. Within six months, 68 professional units with a total of 273 artists had given a total of 2719 performances. 55 of the larger centers in the area were equipped with standard motion picture machines, and 63 similar machines were mounted on 1½ ton trucks which had been fully equipped with electric plants and storage batteries. These portable equipments served the detached units and gave an average of approximately 100 shows every night. A comprehensive educational scheme, including lectures on a wide variety of subjects and a system of regular classes, was quickly developed. Approximately 70,000 copies of the Paris editions of the "New York Herald", "Chicago Tribune" and "London Daily Mail" were distributed daily without charge. German excursion boats, each with a passenger capacity of 500, were requisitioned by the Army. The "Y" supplied each boat with a lecturer who was fully acquainted with the history, geography and legends of the Rhine; it prepared meals from rations furnished by the Army, and provided such additional refreshments as ice cream, cake, etc. Many interesting and instructive trips on the Rhine were made in these boats.

There was drawn up an athletic program which reached practically every man in the Army of Occupation. In addition to mass games, baseball, football, basketball and track events were promoted. The athletic competitions culminated in matches for the championship of the Third Army, the winners competing with the representatives of the troops in France to determine the champions of the A.E.F. An exensive athletic field with a quarter-mile track and a 220 yard straight-away was built on Oberworth Island (called Carnival Island by the Americans), and huts for indoor athletics were constructed. Boxing was encouraged to such an extent that championship bouts were attended by thousands of spectators.

Horse shows and carnivals of various kinds were held during the winter and spring of 1919.[1] Each corps and division had its horse show or carnival, and sometimes held both. In March the Third Army Horse Show occurred at Coblenz, while from April 23 to 27 the Third Army Carnival was in progress there. In a way this carnival was an exposition, as the exhibits covered every branch of the combat services, While there were many amusements, the real value of the carnival lay in its educational benefit; thousands of soldiers attended and secured a comprehensive view of the innumerable parts of that great military machine, the modern army in the field. The Third Army Shows were held on Oberworth Island, which thus received the American designation of "Carnival Island."

Leave areas were opened in Coblenz, Neuwied, Andernach, Trier and Neuenahr, and many sight-seeing trips to points of interest in the American zone were arranged. The Festhalle in Coblenz was requisitioned by the Army and turned over to the Y.M.C.A. After the departure of the units of the Third Army, the Y.M.C.A. made a proportional reduction of its executive and entertainment personnel, and reorganized with a force consisting of the most successful secretaries of the A.E.F. "Y" huts were opened in all remaining garrisons; these were comfortably and attractively equipped, and arrangements were made to provide some form of nightly entertainment for each unit. Professional performances were frequently given at the larger garrisons, while the smallest was provided each night with a new "movie" show.

The center of all "Y" activities was the Festhalle, the municipal

[1] These were G-3 activities, but G-1 and the Y.M.C.A. rendered good service in this connection.

auditorium of Coblenz. The building contained a large lobby with comfortable lounging furniture and capble of seating 1500 people, a restaurant with a capacity of 500, a library stocked with American and English books, and an auditorium with an ordinary seating capacity of 2500. Additional lounging rooms were furnished and writing rooms, game (checkers, etc.) rooms, pool tables and bowling alleys were provided. Two professional entertainment units gave frequent performances, and American moving pictures were shown every night. Amateur companies of soldiers and "Y" personnel were organized and trained. Musical supplies and costumes were furnished and the companies gave performances at all principal garrisons. In addition, many widely known lecturers, musicians and vocalists appeared in the Festhalle during 1920 and 1921.

In attempting to estimate the value of these services to the American troops in the Rhineland, it should be borne in mind that everything, with the exception of the restaurant which was operated on a cost basis, was given without expense to the soldier. No admission fee was charged for any performance given or sponsored by the Y.M.C.A. or by any other welfare organization. Service similar to that in the Festhalle, but on a smaller scale, was given in some 21 other huts in the American area.

As no government funds were available for the operation and equipment of athletic fields, tennis courts, buildings, etc., or for the payment of the salaries of professional athletic directors, the Athletic Department of the Y.M.C.A. took charge of these matters. It defrayed all expenses and furnished all kinds of athletic equipment to the troops, and it was by far the most important factor in the promotion of athletic sports in the A.F.G. An officer, designated as the Athletic Officer (in G-3), co-ordinated the athletic activities in accordance with policies and schedules of military training. These activities comprised a large variety of sports. Records were kept and the names of individuals and their performances were carefully listed. Athletic schedules were planned to conform to the policy of providing athletic recreation for every soldier. This was accomplished by arranging, in conjunction with the program of military training, an extensive mass game program and by providing facilities for a large variety of standard sports so that every officer and man could have an opportunity to engage in his favorite athletic recreation. It was conservatively estimated that 90% of the A.F.G. participated in some form of athletics.

SALVATION ARMY.

The Salvation Army performed valuable and estensive welfare work in the Third Army, covering 15 points among the outlying garrisons and conducting a large club room and canteen in Coblenz. Upon the withdrawal of the units of the Third Army, all Salvation Army huts except those at Selters and Montabaur were closed, and its services were adjusted to meet the needs of the comparatively small number of troops who were to form the A.F.G.

On September 1, 1919, the Engel Hotel in Coblenz was requisitioned for the use of the Salvation Army and after extensive alterations was opened as a soldiers' club room and canteen. Here large and comfortable lounging and writing room was furnished, and reading matter, musical entertainments, social evenings and religious meetings were provided. In December 1919 the Casino in Ehrenbreitstein was requisitioned; a canteen with club rooms was established, together with a theater where moving pictures were shown three times a week.

Religious services were conducted at all central club rooms and at the stations in the outlying districts. The services were well attended, and the instruction given in morality and high ideals was well received.

The beginning of January 1920 found the Salvation Army operating 4 huts in the area. These huts were in Coblenz, Ehrenbreitstein, Montabaur and Selters. The main attraction of the Coblenz hut on Schloss Strasse was the Cafeteria which was complete in every detail, furnishing pies, cakes, doughnuts, etc., made in a modern bakery connected with the hut.

In March 1920 a German café in Coblenz-Lutzel was converted into a soldier's club which provided a writing and reading room, a billiard room and an auditorium for moving pictures.

In April 1920 the Salvation Army equipped rooms in the Disciplinary Barracks and at the Military Prison. There were two rooms at each place, one for the guards and the other for the prisoners. These rooms were furnished with facilities for reading and writing; games, music and moving pictures were provided, and religious services were given twice each week.

Besides the welfare work in the huts, two women Salvation Army officers made daily visits to the Base Hospital, supplying toilet articles, candy, fruit and newspapers.

During May 1920, coincident with the commencement of the rifle practice season, huts were opened at Vielbach, Ransbach, Herschbach, Mogendorf and Nordhofen. Each hut contained an auditorium for moving pictures, entertainments and religious services, and a canteen which was supplied from the Salvation Army bakery in Coblenz. Upon the conclusion of the rifle practice season these huts were discontinued. In 1921 huts were again operated during the target season, one at Selters being substituted for the hut at Nordhofen.

The bakery was established early in 1920 and in the following summer an ice cream plant was installed. Soda fountains and bowling alleys were provided in three huts in Coblenz, Ehrenbreitstein and Coblenz-Lutzel.

In 1920 a silver cup was offered by the Salvation Army to the organization having the best morale report during any one month. This cup after being won three times by the same organization was to become its property. It was finally won by the Medical Detachment Training Center at Neuwied in September 1921. Several chess and checker tournaments were held in Salvation Army huts; many soldiers participated and much interest was aroused.

In addition to these services, soldiers' cash deposits were accepted for safe-keeping, graves in the Romagne Cemetery were decorated, missing relatives and friends were located, destitute American and Allied civilians were helped, and at Christmans time children in and near Coblenz were given useful presents.

YOUNG WOMEN'S CHRISTIAN ASSOCIATION.

No attempt to accomplish welfare work among the troops of the A.F.G. was made by the Y.W.C.A. Its mission was the welfare of women welfare workers and of American and Allied women employed by the A.F.G. and the High Commission. It provided a Hostess House for these women; this house was not only a home for those residing there but was also a social center for other American and Allied women.

The Hostess House was established in Coblenz in February 1919.

It was first located in the Trierischerhof, and accommodations for 36 women were provided. This hotel, later known as the Esplanade, was vacated in December 1919, and the Hostess House was established in the Monopol Hotel building. Here there were accommodations for 75 women; there were a number of rooms available for recreation and entertainments, and the dining room seated 100 people.

Tea was served on Sunday afternoons and was always well attended. There were moving pictures every Monday night, and after December 1919 two dances were held each month.

Until November 1919 the Y.W.C.A. provided a Nurses' Club Secretary at the Base Hospital. A hut was maintained there until August 1919, when it was taken over for medical purposes. The Secretary acted as Hostess and served tea daily; she helped in arranging trips to points of interest and co-operated with the American Library Association in the distribution of books and magazines to the patients.

In February 1920 the Y.W.C.A. furnished three Secretaries at Antwerp to assist, as they did in France, all women arriving or sailing on government transports. Their special work was to look after the wives of soldiers embarking for the United States.

In addition to these activities, the Y.W.C.A. from time to time lodged American and Allied women who made short visits to Coblenz. A very good purpose was thus served, particularly so in the case of women relatives of members of the A.F.G.

AMERICAN LIBRARY ASSOCIATION.

After the dissolution of the Third Army the American Library Association made plans for continuing its library work upon a permanent basis. Before this timethe American Library Association had had its headquarters, together with a small library at the Festhalle, in Coblenz, but the bulk of its work had been carried on through other organizations, in accordance with the general plan for library work in the A.E.F.

The plans for the new system involved the organization in Coblenz of a central library which was to be operated according to recognized library practice.. This library was formally opened on September 18, 1919, at a former German officer's club on Rizza Strasse. The initial collection of books was selected from those provided for the use of the Third Army; it was chosen to meet **practically** any demand for books of study, general information or recreation. In addition to the books for general circulation, a collection of reference books was selected and placed in a special room. The library was later expanded by several shipments of books from the United States, which included the best and newest publications in all lines as well as new and popular fiction. The practice of placing books in the huts ofthe welfare organizations and in company recreation rooms was continued, and branch libraries were established in the larger garrisoned towns.

On December 31, 1920, the A.L.A. withdrew its service in occupied Germany and gave to the War Department all books and property used by it in library service in the A.F.G. As there were go government funds available for carrying on this service, the Y.M.C.A. volunteered to finance the library and to furnish magazines to the Station Hospital and all Y.M.C.A. centers. A central library and reading room was operated in Coblenz, and branch libraries were maintained in each of the "Y" huts. No effort was made to compile statistics pertaining to the use of periodicals and books, but the number of officers, enlisted men and civilians who were constantly in the various libraries showed clearly the need for and the popularity of the form of welfare work furnished by the American Library Association.

AMERICAN RED CROSS.

The American Red Cross began activities in the Army of Occupation in December 1919. Its work was divided into two classes, home service and hospital service. (Prior to this time the American Red Cross had handled 767 home service cases in the Third Army). Each day visits were made to all the wards in the Base Hospital and orders were taken for any purchases patients desired to have made. Many miscellaneous needs were attended to, as for instance the writing of letters for those who were unable to do it for themselves.

Home service work was the connecting link between the soldier in Germany and his home. During 1920 and 1921 the A.R.C. handled over 1400 cases for soldiors of the A.F.G. These covered a variety of subjects, such as bonuses, Liberty Bonds, maintenance, illness, transportation, allowances, allotments, brides, babies, locating families, etc., etc.

The hospital work consisted of the operation of two huts at the Station Hospital, one for the patients andthe other for the enlisted personnel of the hospital. The patients' hut was established primarily for convalescent patients and was abundently supplied with comfortable lounging chairs, games and reading matter. Entertainment was furnished periodically and tea was served daily. The personnel hut, which was operated for the exclusive use of the enlisted personnel on duty at the hospital, contained a stage for theatrical performances, a boxing ring, pool tables, a reading room and a canteen. Moving pictures were shown three times each week, and a vaudeville performance was given once a week. In addition to operating and maintaining these huts, Red Cross Secretaries made daily visits to all hospital wards and endeavored to attend to such of the wants of the men as could not be ministered to by the Medical Corps personnel.

OFFICERS.

The following officers served as Assistant Chief of Staff, G-1, of the Third Army:

Colonel James A. Logan, G.S.
Lieut. Col. Walter B. McCaskey, Inf.
Lieut. Col. James W. Everington, Inf.
Lieut. Col. Stephen C. Reynolds, G.S.

Upon the organization of the A.F.G., Major George M. Peek, C.A.C. (then Lieutenant Colonel, G.S.), was appointed A. C. of S., G-1, and served as such throughout 1920 and 1921.

Majors H. R. Cooper and J. J. Steiner, both of the Corps of Engineers, served at different times as the Welfare Officer of the Third Army. The duties of Welfare Officer of the A.F.G. were performed, respectively by the following officers:

Captain Roy C. Moore, F.A.
Captain W. P. Pinkerton, Inf.
Captain L. S. Gerow, Inf.
(Also Athletic Officer).

CHAPTER NO. 18

MILITARY INTELLIGENCE.

Organisation; Duties - Military Information, Political and Economic Information, Radio Intelligence, G-2 (b), Control of Civilian Circulation, Supply and Distribution of Maps, Censorship, Translations, Publications, The Trier Branch, the Amaroc News, The German Military Commission; Liaison; Officers; Combat Intelligence.

REFERENCE.

General . . . . .A.M.G., Volume I, Chapter 6.

MILITARY INTELLIGENCE.

ORGANIZATION.

The service of Military Intelligence in the A.F.G. was conducted by the Second or Military Intelligence Division of the General Staff (G-2). Due to the fact that it served an army of occupation, the duties of G-2, A.F.G., were more varied than those of a field force. Its organization on July 3, 1919, the date of the dissolution of the Third Army and the adoption of the title of the American Forces in Germany, was the same as that of G-2, Third Army, which was based on that of G-2, G.H.Q., A.E.F. Some of the personnel, because of the signing of the Treaty of Versailles, had been released, but G-2, A.F.G., was largely composed of the same officers and men as G-2, Third Army. The organization, including subdivisions and their specific duties, is set forth below --

A. C. of S., G-2
Assistant G-2
Secretary

G-2 (a-1) Enemy Order of Battle. Military Information.
G-2 (a-2) Political and Economic Information. Review of German press.
G-2 (a-6) Radio Intelligence.
G-2 (b) Secret Service and Counter-Espionage. Control of Civilian Circulation.
G-2 (c) Supply and distribution of maps.
G-2 (d) Censorship-mail, telephone, telegraph, German press, publications and theaters. Permits for meetings.
The Amaroc News. Operation of newspaper.
Trier Branch. Operation of sub-sections of G-2(a-2), (a-6), (b), and (d).

The reduction of the American forces in the Rhineland and the relaxation of restrictions upon the population involved a reduction in the personnel and activities of G-2. From a strength of 37 officers, 7 field clerks and 217 enlisted men on July 3, 1919, by

October 31, 1919, it had been reduced to 22 officers, 5 field clerks and 66 enlisted men; G-2 (c) had been reduced to 2 non-commissioned officers, charged with caring for the large stock of maps and the distribution thereof when ordered, and the G-2 Branch at Trier had been dissolved as the Bezirk of Trier had been turned over to the French.

The coming into force of the Treaty of Versailles on January 10, 1920, caused a lifting of all censorship and restrictions upon civilian circulation, and permitted the release of more officers by G-2. As time went on the reductions in the number of officers authorized for the A.F.G. caused still further reductions in the commissioned personnel of G-2. As its activities had but little specific relation to the number of troops, the former remained about the same during 1920 and 1921. On December 31, 1921, the organization of G-2 was as follows:

A. C. of S., G-2
Assistant G-2
Secretary (Staff Sergeant)
G-2 (a-1)
G-2 (a-2)
G-2 (b)
G-2 (c) - 1 N.C.O. only
Translations.
The Amaroc News.

The strength at the end of 1921 was 6 officers, 54 enlisted men and 9 civilian employees.

DUTIES.

In addition to these outlined above, G-2 was charged with other duties which will be described below. The principal duty of G-2 was to keep the Commanding General and the General Staff of the Army of Occupation thoroughly informed upon conditions in the American area and to secure accurate military, political and economic information of Germany, primarily for the Headquarters at Coblenz and secondarily for G-2 at G.H.Q., the American Delegation to the Peace Conference, the Military Intelligence Division at Washington, the Intelligence Services of the associated forces on the Rhine, and American military attaches in Europe. Various other duties were assigned to G-2, but all were subsidiary to the principal duty mentioned.

MILITARY INFORMATION.

G-2(a-1) was charged with the keeping the enemy order of battle; after the signing of the Peace Treaty, the war-time phraseology was abandoned and the general term "military information" was designated as the duty of this subdivision. Until the old German Army was completely demobilized, close touch was kept with the progress of its dissolution; after that G-2(a-1) kept in similar touch with the numbers, composition and stations of the Reichsheer (National Army), the title given the reorganized forces, and with the Security Police and the various "self-defense" organizations. Through liaison with the American military attache at Warsaw and other sources, good information was secured of the Bolshevist army and of the operations between it and the Polish army in the summer of 1920. After the spring of 1921, as the Reichsheer had been located in its permanent stations, the keeping of accurate records became a matter of little difficulty. G-2(a-1) then devoted most of its energies to securing military information of historical value and copies of new military regulations issued by the Reichswehr (National Defense) Ministry; these were translated, edited and sent to the Military Intelligence Division in Washington.

All military information, including most of that of an historical nature was collated, recorded, and published in periodical bulletins.

Political and Economic Information.

G-2 (a-2) collated and recorded all political and economic information of Germany that came to G-2 from various sources. A resume of this information was published in bulletin form. A daily review of the German press was also published by this subdivision and distributed to Headquarters and the larger units.

Radio Intelligence.

In the early days of the occupation, the duties of G-2 (a-6) were of considerable importance and were somewhat analogous to those of war times. They became less and less so as the demobilization of the German Army progressed, and it was decided to use the personnel for the preparation and publication of a daily radio news bulletin. The Signal Corps intercepted news dispatches from the Continent of Europe and the United States, and turned them over to G-2 (a-6). These dispatches came from stations at Nauen (near Berlin), Lyons (France), England, Annapolis (Maryland), Copenhagen, Moscow, Stockholm and Constantinople -- dispatches sent out by the first four stations were generally used, the others only occassionally. They came in German, French and En lish, and were sent in the concise form and abbreviated words common to all press dispatches. G-2 (a-6), selected the most interesting items, translated them when necessary, put them into conventional English, and published them daily in mimeographed sheets which were distributed to Headquarters, A.F.G., to the headquarters of the larger units, and to the High Commission.

G-2 (a-6) was disbanded in the spring of 1921; thereafter the Signal Corps furnished the radio interceptions to the Amaroc News which published the more important items that were sent in English.

G-2 (b)

This sub-division was at first composed of several officers and of a number of Intelligence Police. In 1921 the commissioned personnel was reduced to one and at the end of that year the Intelligence Police had been reduced to 14. Among its duties were the following:

> Observation of the civil population, including transients, and deportation of undesirables.
> Attending and reporting upon political and other meetings of interest.
> Securing information of German Army and furnishing it to G-2 (a-1).
> Securing political and economic information and furnishing it to G-2 (a-2).
> Indirect prevention of strikes by the securing of early information of industrial disputes.[1]
> Detecting and thwarting improper propaganda.
> Counter—espionage.
> Liaison and co-operation with similar agencies of associated armies.
> Liaison and co-operation with Provost Marshal, A.F.G.
> Special missions.

Control of Civilian Circulation.[2]

This control was exercised partly through operatives of G-2 (b) but mainly through the operation of a pass bureau. Technically the sub-division known as "Civilian Circulation" was a part of G-2 (b), actually it was not, although close liaison and co-operation prevailed between the two. The former operated a pass bureau to which all applicants (except those at Trier) to leave the American

[1]Many strikes were prevented through liaison and co-operation between G-2 and Civil Affairs.

[2]A.M.G., Volume I, Pages 141-143.

area reported. The restrictions upon the circulation of the German inhabitants were gradually relaxed during the last half of 1919 until on January 10, 1920, they were entirely removed, and the Civilian Circulation sub-division was disbanded.

Two non-commissioned officers were retained to operate a pass bureau under the supervision of the Secretary. This bureau attended to the securing of passes for members of the A.F.G. to enter or pass through Germany. Its functions gradually extended until it became a bureau of information relative to passports and travel in general.

This bureau acted also as a medium for the securing of domestic servants by the members of the A.F.G. and the Allied community, and in this way served a beneficial purpose.[1] This function was given up in the autumn of 1921, the German municipal employment agency being used thereafter.

Supply and Distribution of Maps.

In the Third Army G-2 (c) was composed of several offices and a large number of men - all from the 29th Engineers. It possessed a large and well equipped plant and was divided into map, drafting, printing and lithographic, photographic and topographic departments. It performed the printing for the Third Army and supplied all maps used by that force. During July 1919 the functions of G-2 (c), except those of the map department, were ended, its plant was dismantled and shipped to Washington, and most of the

[1]In the latter half of 1919 house servants and cooks were paid the equivalent of $1.75 and $3.50, respectively, per month. In February 1920, when the mark had suddenly dropped to one cent in value, in one household there were three servants whose total wages were the equivalent of $1.90. Wages rose gradually during 1920 and 1921, but not in proportion to the depreciation of the German currency. At the end of 1921, in the same household, the total wages of three servants amounted to $5.00.

personnel were returned to the United States.

A very large stock of maps was left and the remaining functions of G-2 (c), now reduced to two men, were the care and distribution of maps under the supervision of the Secretary. A supply of excellent maps of the American area and vicinity, large enough to meet the needs of the A.F.G. in the event of active operations, was purchased in Berlin. This stock of maps was kept constantly on hand, maps used during maneuvers being immediately replaced.

Censorship.

Next to G-2 (c), G-2 (d) was the largest sub-section. Its main duty was the censorship of mail, telephone conversations and telegrams, and the publication of a daily resume of the information gained thereby.[1] Some time after the signing of the Treaty of Peace the armies of occupation agreed upon a relaxation of the censorship, and on August 25, 1919, postal telephone and telegraph matters were permitted to resume their normal courses. Censorship continued, but it was performed by sondage - in general a system of censoring mail and telegrams coming from particular sources and listening in on particular telephone lines (a daily change being made), together with a weekly trip to some town in the American zone, where the censors would operate for a day, returning to Coblenz at night. This permitted the release of the larger portion of the censorship personnel.

There was no pre-censorship of the press or other publications.[2]

[1] A.M.G. Volume I, Pages 156 - 157.

[2] Ibid, Page 155.

Publishers and book sellers were held responsible that their journals or other publications contained nothing offensive or injurious to the American authorities. Newspapers violating the American regulations were punished by suspension for one or several days,[3] and on several occasions book sellers offending were fined.

Theaters and moving pictures were visited from time to time by personnel of G-2 (d), but it soon became apparent that there was little need for censorship of even so mild a type. The practice was continued, however; in a few instances the showing of certain films was forbidden, but no occasion arose to punish theater proprietors because of plays offensive to the Americans. They were very careful and when the slightest doubt as to a play arose in their minds, they immediately put the case before G-2.[1]

Permits for meetings were issued by G-2 (b).[2] The meetings were covered by operatives of G-2 (b); a daily resume of the reports rendered by these operatives was made and copies were furnished the staff officers concerned. Permits for meetings were not required after January 10, 1920, but political and labor union meetings were covered by G-2 (b) throughout 1920 and 1921.

G-2 (d) was disbanded in February 1920.

Translations.

[3]This action was taken in several instances.

[1]The local opera company was to give "Madam Butterfly", the hero of which is an American naval officer. The proprietor, fearing that the portrayal of an American officer who was known to have had illicit relations with a Japanese woman might offend the American authorities, offered to alter the part and cause the hero to be a citizen of an imaginary country. Permission was given to produce the opera without alteration of any kind.

[2]A.M.G., Volume I, Page 151 - 154.

Early in 1920 a special sub-division was organized to take over the constantly increasing amount of translating required by the Headquarters. An officer, and later adischarged emergency officer, was placed in charge and several civilian translators were employed. In addition to making translations for other Staff Divisions and Services, this sub-division translated a number of German military books of a technical nature and a great many articles of military and political nature. The bulk of these translations was from the German, but many were in French, occasionally a paper in some less known language was received, and it was always possible to find some one in the A.F.G. who could make the necessary translation.

An average of 75,000 words were translated each month, in addition to the routine translations for G-2.

Publications.

The following routine publications were issued by G-2:

Daily

- Review of German Press.
- Report of Meetings - by G-2 (b)
- Excerpts from Reports of G-2 (b) Operatives
- Radio Press (discontinued April 1921)
- Items of Censorship Information (discontinued February 1920)

Weekly

- Bulletin of Military Information.
- Bulletin of Political and Economic Information
- Circular Letter (to Military Attaches)
- Resume of Censorship Information. (discontinued January 1920)
- Digest of G-2 (b) Information.

Semi-Monthly

- Estimate of the Political Situation.

Reports on special subjects, such as the Ruhr Revolt and Upper Silesia, were made as the occasions arose, and numbers of "monograpn reports" were made to the Military Intelligence Division at Washington.

The Trier Branch.

The branch at Trier had an organization somewhat similar to that of G-2 at Coblenz, but on a smaller scale. It served Advance G.H.Q., which was located in that city, as well as the Third Army. It gathered political and economic information, published a daily bulletin of radio news, conducted a G-2 (b) service, a pass bureau and a censorship of the mail, telephone and telegraph, publications and theaters, issued permits for meetings and covered the latter, and published a daily review of the German press. The Trier Branch was disbanded in September 1919.

The Amaroc News.

This unique newspaper was established in April 1919.[1] It was placed under the broad supervision and control of G-2 and became a special sub-division thereof.

The personnel was composed almost entirely of officers and enlisted men, including a correspondent in Paris. It was "started on a shoestring", and other than the pay and allowances of its personnel, no Government funds were involved in its financial management.

The paper was a success from the beginning and after the first month there were no financial difficulties. Its circulation grew rapidly from 10,000 on April 21, 1919, the date of the first issue, to the maximum of 68,000 in June 1919. By August

[1]The title of the paper was chosen by Major General Dickman, commanding the Third Army, from a number of suggestions made to him. The work "Amaroc" was composed of the first syllables of the words "American Army (of) Occupation" The word was adopted by s[illegible] for a number of [illegible]d., the American Forces in the Rhineland.

it had declined to 30,000 and by January 1920 to 6500. During the latter part of 1920 the circulation was gradually built up to 8000, and despite the gradually decreasing strength of the A.F.G., the circulation was maintained at an average of 7500 during 1921.

The Amaroc News would have experienced serious financial difficulties during the summer of 1919 but for the co-operation of the Y.M.C.A. and the Knights of Columbus. The units of the former Third Army were departing for the United States and while the announced decision to cease publication had been rescinded, the paid circulation fell to an alarmingly low point. The two welfare organizations contracted, on July 15, for practically the entire output and distributed the paper free of charge to officers and enlisted men. This practice was continued until the end of October when, at the behest of G-2 and the editor of the paper, the two welfare organizations withdrew their contract. After that the paper was conducted on a purely business basis, and it remained self-sustaining throughout 1920 and 1921. Its large circulation during the first three months and the co-operation of the Y.M.C.A. and the Knights of Columbus had resulted in the accumulation of a large surplus which remained on hand at the end of 1921.

The accounts were kept in German currency, the depreciation of which combined with the increased costs of material and labor caused the price of the paper to be altered from time to time. In the beginning the price was 50 pfennigs per copy. In November 1919 it was increased to 75 pfennigs, and in December to one mark. In June 1920 the price was reduced to 50 pfennigs. In November one mark was charged for the Sunday edition, but no other increase was made until November 1921 when the price was fixed at one mark for all editions. The sudden depreciation of German currency in November 1921 caused the price to be raised to two marks.

At first the paper contained no advertising, but when the circulation decreased advertisements were sought and they became an important source of income.

The Amaroc News was a four page newspaper with a Sunday edition of eight pages, including a comic section of four pages. Several times a month, usually in the middle of the week, a six page edition was published in order to accomodate the large amount of advertising which had been secured.

The printing was done by a German firm which also printed a Coblenz newspaper. Some German mechanics were employed by the Amaroc News, but with that exception practically all of the work of the paper, including linotyping, was done by the American personnel.

The front page of the paper was devoted to news of the world which was secured by telephone and telegraph from Paris and from Signal Corps radio interceptions. Until July 1920 the Amaroc News had a special correspondent in Paris, but during that month a contract was made with the Paris edition of the Chicago Tribune and the correspondent was discharged. Thereafter the paper had the benefit of the Chicago Tribune news service, and its front page contained the latest and most important items, those from the United States predominating.

The paper was the recipient of many favors from news syndicates and press correspondents who furnished it with good news stories from time to time and enabled it to make free use of material such as

"Mutt and Jeff" and "Bringing up Father" for its comic supplements.

Local news was secured in the ordinary manner by enlisted reporters, and also from German newspapers with which the Amaroc News exchanged information. A special feature of the Sunday edition was the "Organization News" page which was devoted to items concerning the various units, secured from correspondents appointed by the respective organizations.

Extras were issued on several occasions such as the signing of the Treaty of Versailles, the American presidential election, etc. A number of pictorial supplements were published, the most elaborate of which, issued on Christmas day 1921, consisted of 12 seven column pages of high quality book paper.

The paper was distributed in Coblenz and to the troops in the outlying districts through an agency with which a contract had been made; deliveried outside of Coblenz were made by motor transportation which carried the newsboys to the various headquarters. It was delivered to the troops every morning during maneuvers and target practice. The list of mail subscribers was quite large; the majority of these were in the United States, but there were some in almost every country in Europe.

The Amaroc News began its career with a staff of 2 officers and 37 enlisted men, all of whom were trained newspaper men. During the spring and summer of 1919 there were a number of changes in the staff, which made a reorganization necessary. As reorganized it functioned until the autumn of 1920 when the return to the United States of almost all the personnel caused another reorganization. The War Department had been requested to enlist newspaper men for the Amaroc News, and several had arrived by winter. But most of the reorganized staff were comparatively inexperienced, and the paper entered upon a difficult period. Despite many obstacles, however, it attained its maximum efficiency during 1921. At the end of the year, in consequence of the reduction of the A.F.G., a third reorganization was impending. The average enlisted strength during 1920 and 1921 was 24.

At the outset 1st Lieutenant Theodore E. Damm was business manager and 1st Lieutenant William W. Corcoran was managing editor. After a few weeks the latter was succeeded by 2nd Lieutenant Guy C. Stafford, who in June took entire charge of the paper in consequence of the departure for the United States of Lieutenant Damm. Lieutenant Stafford served as editor and manager until October 1920 when he was relieved by Major Paul J. Mueller, Infantry, who remained in charge throughout 1921.

The Amaroc News was unique in Army annals, and its efficiency reflected much credit upon the officers and men who composed its staff. With the exception of December 26, 1920, it was issued daily and Sundays from April 21, 1919, to the end of 1921, and it was planned to continue its publication indefinitely.

The paper was a great aid to the morale of the troops and it was keenly appreciated by the American and British Community. It was a newspaper and not an official gazette, and its columns were kept free from propaganda of any kind. Its attitude toward the Germans was one of dignity and no abuse or ridicule was permitted. The broad lines of its editorial and news policy were prescribed by G-2, but otherwise it was conducted practically entirely by the editor and manager and without any interference as to details. It was ably conducted, especially during 1921, and it served a very good purpose.

Copies of every issue were sent to the Adjutant General of the Army, to the Military Intelligence Division at Washington, and to the library of Congress. Every anniversary edition contained a history of the paper; these who desire a more detailed account of the Amaroc News can secure it by consulting the files mentioned above.

The German Military Commission.

The Deutsche Militär Kommission, or German Military Commission, was the outgrowth of the early necessity of liaison with the German forces and of the provisions of the Armistice Agreement concerning the delivery of German war material. In each of the towns of Coblenz, Cologne, and Mainz there was, from the beginning of the occupation, a German staff officer known as the (Coblenz) Bridgehead Staff Officer. His principal duties were the delivery of war material under the Armistice Agreement and the maintenance of liaison between the headquarters of the troops occupying the bridgehead and the Commander of the German troops in the neutral zone opposite that bridgehead. A sufficient personnel for the performance of these duties was allowed, and in Coblenz the office was given the designation set forth above, usually abreviated to D.M.K.

The D.M.K. functioned under the immediate supervision and control of G-2 and all communication with the American Headquarters had to be conducted through that office. In this way to D.M.K. was under constant observation and a close check upon its activities was kept.

The delivery of the German War material (guns, machine guns, aeroplanes and motor transportation) was not completed until the the late summer of 1919, at which time a portion of the personnel of the D. M. K. was transferred to other duties and stations.

The liaison with the German Commander of Sector III of the neutral zone, which was opposite the Coblenz bridgehead, continued until the evacuation of the neutral zone in September 1920. The first matters requiring attention were the accurate locating of the perimeter of the bridgehead and the placing thereon of the necessary control posts. Numerous questions, such as the entering of the neutral zone by American soldiers (of which there were many instances) and the sending of American patrols to arrest the offenders, the required inspections of the German troops in the neutral zone, claims for damage, arrangements for passes, etc., arose from time to time; it was found that the system of communication through the D.M.K. rather than through the Armistice Commission at Spa worked very well and saved much time.

As time went on it became apparent that the D.M.K. could be of considerable use to the American Headquarters in matters that originally had not been within its purview. It had always been charged with the supervision and disciplinary control of all members of the German Army who were officially present in the American area. Numerous deserters came into the American zone, and the D.M.K. was directed to assume control of them in accordance with the policy adopted by the Americans. This policy was that all German deserters in the area were to be demobilized at once; only in case of proof that crimes such as murder, robbery or serious assault had been committed in unoccupied Germany by the deserters, could they be taken there for trial.

On the other hand there were not a few American deserters, absentees or criminals who fled to unoccupied Germany and were there apprehended and brought back to the American area through the medium of the D.M.K.

Through the D.M.K. passes premitting travel into unoccupied Germany were secured, and all requests from German military or naval personnel to enter the American area came to G-2 through that office. The former activity was expanded to include passes for American civilians, and was later turned over to a passport bureau established in Coblenz by the German Foreign Office. The intervention of the D.M.K. including the furnishing of requests that the journeys through Germany of American personnel be facilitated, proved to be of much benefit to those whose duty took them there.

Liaison with the German War Office was conducted by the D.M.K., and through this liaison much information of historical as well as present value was secured and forwarded by G-2 to Washington. Many inquiries as to American prisoners of war, the location of American dead and the circumstances of their deaths came to G-2; in almost every case the desired information was secured through the D.M.K. Reciprocally, similar requests of the German War Office were received from the D.M.K. and forwarded to Washington.

The matters handled through the D.M.K. came to be numerous and among them were the following:

Repatriation of German and Czecho-Slovak prisoners of war from Siberia.

Repatriation of German war prisoners from France.

Necessary communications between the American Headquarters and civil or military authorities in unoccupied Germany.

Passage through Germany en route to Russia, of American Red Cross units.

After the coming into existence of the Rhineland High Commission and the appointment of aGerman Commissioner, the activities of the D.M.K. gradually decreased. Its chief was attached to the German Commissioner as the representative of the German War Office for the whole occupied territory. At the end of 1921 the D.M.K. was still functioning, although with a reduced scope, and it continued to be of much service to the American Headquarters in various ways.

At the beginning of the occupation the D.M.K. consisted of the Chief (an officer of the General Staff), 8 other officers and 19 non-commissioned officers and men. At the end of 1921 the personnel comprised one major, one lieutenant, 4 non-commissioned officers, 2 enlisted men and 3 civilians.

The first "Staff Officer of the Coblenz Bridgehead" was Captain von Sybel, Chief of Staff of the 76th Reserve Divison. He was relieved on January 1, 1919 by Major Eric Kuehlenthal who remained at the head of the D.M.K. throughout the next three years. The latter officer received his commission in 1899 and in 1913, as a captain, was detailed on the Great General Staff. During the World War he was Chief of Staff of several divisions, serving in both the eastern and western theatersof operations. He received ten decorations and was promoted to the rank of major in July 1918.

As head of the D.M.K. Major Kuehlenthal displayed marked tact, discretion and ability; his situation was always delicate but was handled with such skill as to win the commendation of the American Commander. After three years of dealingswith the American Staff, Major Kuehlenthal enjoyed their confidence in his integrity and straight dealing and their appreciation of the many services rendered them both officially and personally.

On the other hand two of the officers who served with the D.M.K. conducted themselves so as to require their removal from the American area. Captain Schultz-Bachhusen displayed a lack of tact and a spirit of arrogance; he violated orders issued by G-2 for the guidance of members of the D.M.K. and as a matter of discipline his chief was required to relieve him from duty and to send him out of the American zone.

Captain Quensell was of a different type, but while intoxicated he entered into an altercation with American military policemen who placed him in confinement. He was tried by a military court and sentenced to fine and imprisonment. He was released before the expiration of his term, and was required to leave the American area at once.

None of the other officers got into any difficulty while serving with the D.M.K. Those who came most in contact with the Americans were:

Captain von Schuckmann.
1st. Lieut. Bartholdy.
1st. Lieut. Gescher.
1st. Lieut. Melmer.
2nd Lieut. von Schweppenburg.
2nd Lieut. Frischmann.

Liaison.

G-2 was in constant liaison with the Military Intelligence Division at Washington, sending them copies of all reports except those of minor local interest. A number of special reports were made to it, while German military books and articles of historical value were translated and forwarded there.

Liaison with all American military attaches in Europe, including those in Turkey and Egypt, was established in the summer of 1919 and was maintained continuously thereafter. A constant interchange of information was made, and in addition G-2 forwarded them a weekly circular letter which included condensed information of Germany and items of Army interest. In the winter of 1920, G-2 attended the conference of military attaches in Rome. Similar conferences were held at Coblenz in July 1920 and June 1921.

Through the American liaison officers with associated armies of occupation, who functioned under G-3, G-2 secured information of various kinds. Colonel Le Vert Coleman, C.A.C., stationed at the French Headquarters at Mainz, secured an extended study of the new organization of the French army and the methods adopted as a result of the experience gained during the World War; this series of reports was furnished to G-2 and by that office forwarded to Washington.

Constant liaison with the Intelligence Services of associated armies was maintained. There was a continuous interchange of information, conferences were held from time to time, and cordial relations prevailed throughout 1919, 1920 and 1921.

OFFICERS.

Colonel R. H. Williams, G.S. was A. C. of S., G-2, of the Third Army. When the A.F.G. came into existence he was relieved and ordered to the United States. Major (then Lieutenant Colonel) P. H. Bagby, G.S., was then appointed A. C. of S., G-2 of the A.F.G., and served in that capacity throughout 1920 and 1921.

Major (afterwards Lieutenant Colonel) Henry Hossfeld, Infantry, joined G-2 in the late summer of 1919, and served as Assistant, G-2 until his promotion late in 1921. He was then assigned to the office of the Chief of Staff, but was attached to G-2 in a special capacity.

The following officers, among others, served in G-2 during some portion of the period July 1919 - December 1921:

Major E. W. Savage, Inf.
Captain H. K. Coulter, Inf.

G-2 (a-1)
*Capt. B.B. McMahon, Inf.
Capt. A. E. Dedicke, Inf.
1st. Lieut. T. R. Taber, Cav.

G-2 (a-2)
Lieut. Col. O. Seebach, Inf.
Maj. O. H. Fernbach, C. of I.
Maj. G. E. Arneman, F.A.
Capt. J. W. Bollenbeck, Inf.
Capt. L. E. Schoonmaker, C.A.C.

G-2 (a-6)
Capt. H. A. Berthold, C.A.C.
Capt. E. M. Nourse, M.T.C.
*1st. Lieut. C. W. Bird, C. of I.

G-2(b)
Major A. M. Scully, Inf.
*Capt. H. E. Osann, Inf.
Capt. H. C. Minuth, Cav.
Capt. W. E. Beitz, F.A.
1st Lieut. P. G. Balcar, Inf.

Civilian Circulation.
1st Lieut. M. J. Stern, Inf.
1st Lieut. G. H. Ramm, Inf.
2nd Lieut. A. Florin, Inf.

G-2(c)
Major H. R. Richards, C. of E.
Capt. R. N. F. Townsend, C. of E.
Capt. A. R. Stroup, C. of E.
Capt. J. G. Goedike, C. of E.

G-2(d)
Capt. F. L. Berryhill, Inf.
Capt. F. M. Distelhorst, Inf.
1st Lieut. A. H. Aldridge, Inf.
1st Lieut. J. C. Dieckhoff, T.C.

Translations.
*1st Lieut. C. W. Bird, O.R.C.

Trier Branch.
Major J. W. Downer, F.A.

*Officers who served in G-2 for more than 2 years.

Major E. F. Rice, G.S.
Capt. E. Cordes, Inf.
1st Lieut. P. G. Balcar, Inf.

(Staff Sergeant David B. Dugan served as Secretary of G-2 for over a year)

COMBAT INTELLIGENCE.

Far from being allowed to lapse, the service of Combat Intelligence in the troops of the A.F.G. was brought to a high standard. There was an Intelligence Section in each regiment and separate unit, including the Air Service Detachment.

The 8th Infantry, being then at war strength, organized an Intelligence Section in accordance with Intelligence Regulations in the late summer of 1919 and by the opening of the field training season in 1920 had achieved marked efficiency in this line. Due to many causes, including the omission of Intelligence Sections from their tables of organization, the 5th and 50th Infantry regiments did not reach their maximum efficiency until 1921. By that time, throughout the A.F.G. the service of Combat Intelligence was functioning smoothly and efficiently, and in the autumn maneuvers of 1921 the efficient service rendered the various units by the Intelligence Sections brought forth commendation from all sides.

In the maneuvers of 1920 and 1921 the troops of the A.F.G. were provisionally organized as a small division and the service of Combat Intelligence operated along similar lines. G-2 furnished a Division Intelligence Section at reduced strength and there were Intelligence Sections in the brigades as well as in the lower units. In the maneuvers of the latter year the service reached its maximum efficiency and showed clearly that in Combat Intelligence as in other branches of training the A.F.G. were ready to take the field.

Policies and programs of training were formulated by G-2 and transmitted to the troops through G-3; in this way there was but one doctrine of Combat Intelligence in the A.F.G. The duty of supervision of the training was performed by the A C. of S., G-2, who made frequent visits to the various units and kept constantly in touch with Combat Intelligance matters throughout the command.

CHAPTER NO. 19

MILITARY TRAINING AND OTHER ACTIVITIES OF THE THIRD DIVISION, GENERAL STAFF.

Standards Required; Military Training - Tactical (Intensive Training, Tests, Maneuvers, Results), Plan of Defense, Competitions, Equitation; Athletics - Interallied Competitions; Liaison Officers with Associated Armies; Officers.

MILITARY TRAINING AND OTHER ACTIVITIES

OF THE

THIRD DIVISION, GENERAL STAFF

Soon after the American Forces in Germany came into being it became apparent that American troops would remain in Germany for a more or less indefinite length of time. In the matter of military training - of general fitness for active service - it was necessary that they be thoroughly efficient. They would be the cynosure of all military eyes in Europe, and the effect upon European political conditions of their presence on the Rhine would probably be far greater then their numbers would seem to warrant. For all these reasons it was required that the A.F.G. be worthy representatives of the American people.

The Commanding General directed the immediate institution of a policy of training that not only would cause the A.F.G. to be a thoroughly efficient combat unit in general, but also would insure the maintenance of high standards in such details as discipline, military bearing and appearance, military ceremonies, equitation and athletics. The Third Division of the General Staff (G-3) was charged with the formulation of policies governing those matters and with the supervision and co-ordination of all activities connected therewith. [1]

MILITARY TRAINING

Tactical

[1]During 1920 and 1921 it was reported by American Military Attaches and others that throughout Europe the A.F.G. had acquired a high reputation for military efficiency. Foreign officers who visited Coblenz were almost lavish in their praise.

A thorough and exhaustive program of intensive training was drawn up by G-3 and issued to the troops. It began with the training of the individual soldiers, extended progressively through all phases of the training of tactical units, and culminated in autumn maneuvers which lasted several weeks and in which the entire A.F.G. participated.

The conditions were ideal, due largely to the powers possessed by an occupying force. A great deal of the country that was comprised within the American area is rolling, of varied type (open, wooded, etc.), and is practically fenceless. When appropriate terrain for tactical exercises or maneuvers was desired, it was only necessary to select the ground and to notify the proper German authorities. - The Americans then secured unmolested possession of the tract chosen. The exercises and maneuvers were held in the autumn after all crops had been harvested. The damages, which usually amounted to little, were paid by the German Government and included in the costs of occupation. A vast amount of small arms and other varieties of ammunition, much of which would deteriorate if not used within a short time, had been left behind by the Third Army. Hence it became feasible to have combat exercises and to use live ammunition on a scale that would have been impossible in the United States. Altogether the superior advantages afforded the A.F.G. were such as to cause the command to be a valuable training school for officers and men. Not the least of those advantages was the fact that the strength of the various units was much greater than in the case of similar units at home; the 8th Infantry and the 2nd Battalion of the 6th Field Artillery were kept at war strength until the latter part of 1920, while until a year later all units had strengths well above those that were possible in the United States.

An important phase of the annual training program comprised tests for each infantry battalion, during which an extensive tactical problem had to be solved. The solution required the tactical employment of an infantry battalion, including machine guns, trench mortars and one-pounders, with a battery of artillery, all of which were under the orders of the battalion commander. Infantry planes (in 1921) were used, and the conditions were made as nearly similar to those of actual combat as was possible. The operation culminated in the actual firing of live ammunition in an attack upon a prepared position. The problems were carefully prepared, exhaustive critiques were held at the conclusion of each exercise, and battalions were rated as satisfactory or unsatisfactory. A keen spirit of competition was aroused, and the general results of this phase of training were excellent.

In March, 1920 practically the entire command was concentrated in the vicinity of Ransbach. The troops engaged in maneuvers for about a week, after which the assembled units were inspected and reviewed by the Commanding General. From August 30 to September 16 the troops engaged in brigade and division maneuvers, which were preceded by roundabout concentration marches during which the various units had tactical exercises of a minor nature. The maneuver area was in the country west of Mayen. The maneuvers concluded with the solution of division problems by the "A.F.G. Division"; these included the assembly of the division, the relief of an imaginary division in front line, an attack on a (simulated) prepared position occupied by an outlined enemy, and the exploitation of an initial success. The battalion tests, mentioned previously, were conducted between November 1 and November 15, after which the field training season ended.

During 1921 the field training was conducted along similar lines. The battalion tests were more extensive and were held during August. During September brigade and division maneuvers, lasting three weeks, were held in the Kaisersesch-Bassenheim-Kruft-Laacher See-Niedermendig area. From the first week the command was divided into two reinforced brigades, and brigade commanders held such inter-brigade maneuvers as they desired. Brigade-versus-brigade maneuvers were held during the second week; these involved, for each brigade in turn, the gaining of contact with the enemy, dispositions for attack, an attack, a withdrawal, and the occupation of a defensive position. The third week was given over to maneuvers by the "A.F.G. Division". The first phase required the assembly, by night and in a wooded area, of the entire command; a division sector was then taken over and outposts were established. As part of an attacking army the division advanced over difficult terrain against an outlined enemy - the advance including the crossing of an unfordable stream. Next came the temporary occupation and organization of a defensive position pending a further advance. The final maneuver involved an attack on a prepared position which had been organized in detail by engineer troops, including a few actual trenches and the extensive use of wire entanglements. Live ammunition was used by all arms, overhead fire by artillery and machine guns was employed, and explosive targets added to the general effect. Both friendly and hostile airplanes were used during this problem, as well as during the whole maneuver period. General Pershing was present during the final problem, and two days afterward he reviewed the entire A.F.G. at Weissenthurm. The efficiency displayed by the A.F.G. during these maneuvers indicated that the command was well prepared for active service. The appearance of the troops at the final review was splendid, and it demonstrated the excellent results that can be achieved through a system of frequent inspections and reviews by brigade commanders, supplemented by monthly inspections and reviews of brigades by the Commanding General.

Plan of Defense.

In the summer of 1919 a plan of defense was prepared. This plan was modified from time to time, because of the changing situation with regard to the number of troops available, as well as in accordance with modifications of the general plan of defense of the associated armies of occupation. When circumstances required, conferences were held at Mainz under the supervision of the Commanding General (Degoutte) of the Allied Armies of Occupation, and during these the co-ordination of the plans of the four associated armies was affected.

Competitions.

Inter-allied competitions will be discussed later, but it may be said here that those held in connection with horse shows, as well as similar A.F.G. competitions - among artillery sections, machine gun sections, vehicles, animals, motor transportation, etc. - had a marked influence in bringing the appearance of animals, animal and motor transportation, and equipage to a state of perfection never before attained in the United States Army.

Equitation.

Each year from October to the end of March classes in equitation were held. Every officer in the A.F.G., unless a thoroughly competent horseman or excused for physical reasons, was required to undergo this instruction. The results achieved were highly satisfactory under all the circumstances. It is probable that the officers of the A.F.G. comprised a larger proportion of capable riders than ever before had been the case in an American force which included officers of every branch of the service.

ATHLETICS

Every variety of athletics and sport was promoted. The Commanding General realized that troops required to undergo such intensive military training would need healthy diversion, not only to add to their contentment but also to provide for their physical health. With this in mind, the individual was required to take certain physical exercises in conjunction with his military training, while his participation in general athletics and sports was constantly encouraged. In addition, interest was aroused and stimulated by all sorts of athletic competitions, with the other armies of occupation as well as among the units of the A.F.G.

In the promotion of athletics and sports the Y.M.C.A., with its corps of athletic experts and executives, was of invaluable assistance, and in fact was largely responsible for the excellent results achieved. This welfare association expended large sums of money in the promotion of athletics and sports, including the provision of prizes for the many competitions that were held.[1] The athletic field, constructed on Carnival Island by the Y. M. C. A., was exceptionally good. It included a cinder track, baseball and football fields, tennis courts, a horse show ring and a jumping course; all

[1] Perhaps the most interesting of the individual competitions was that designated as the "Best Yank Competition" (held every year). This included a series of athletic tests of every variety, the object being to bring forth, by a process of elimination, from among all the soldiers of the A.F.G., a champion who would be known as the "Best Yank".

these were symmetrically and conveniently arranged, and stands, booths, locker rooms and score boards were provided. The Y. M. C. A. also assisted in the construction of the Country Club, at the polo field on the heights south of the city.[2]

Among the various forms of athletics and sports promoted in the A.F.G. were basketball, football, soccer, baseball, boxing, tennis, polo, golf, mass games, outdoor and indoor meets, handball, bowling, and gymnasium classes. In practically all of these sports leagues were formed and, particularly in baseball and football, great enthusiasm was aroused.

Inter-Allied Competitions

There were many competitions among the four armies of occupation, among the more notable ones were the following:

> An Inter-allied military competition (rifle, automatic rifle, machine gun and pistol) was held at Herschbach (American area) in 1920. The American team was first in every competition except that with machine guns, in which it took second place.
>
> In a similar competition at Aix-la-Chapelle (Belgian area) in July 1921, the Americans again won every event except the machine gun competition.
>
> The French-American indoor track and boxing meet, held at Coblenz in January 1920. The French entered over 50 athletes, and the A.F.G. won all but two events.
>
> Inter-allied athletic meet, held at Mainz (French area) in March 1920. The A.F.G. entered 16 events, and won 13 firsts, 2 seconds, and 1 third place.
>
> Inter-allied athletic meet at Mainz in June 1921. The A.F.G. finished as a close second. The Americans won in basketball and soccer, while Private Paul (8th Infantry) won the "Coq du Gaulois" which had been offered by General Degoutte for the best all-around athlete. (This soldier was the "Best Yank" in 1921).
>
> The Inter-allied athletic competition, held at

[2] Volume II, Chapter 17.

Coblenz in July 1921, was easily won by the Americans.

Inter-allied horse show, at Coblenz in July 1921. The Americans won the officers' jumping event (56 entries) and the enlisted men's jumping event (40 entries), while they took second place in the jumping in pairs.

During 1920 and 1921 a jumping team engaged in competitions with teams from the other armies of occupation, and also participated in some of the highest class competitions in various large cities of Europe. The team was very successful, and the experience gained was of considerable value.

Polo was a constant recreation from May to October. During 1920 a polo team was sent to England where it did very well against strong English teams. By 1921 a considerable number of polo players had been developed, and there were five regualr teams in the A.F.G. In a tournament at Coblenz in October of that year, the American team was victorious over a strong British team (14th Hussars).

A small string of race horses was developed. In contests with the British and French the Americans scored a fair percentage of victories, in both flat races and steeplechases. A race course was constructed around the polo field, and monthly race meets were held there during the 1921 season.

A hunt club was formed in 1921, and hounds were brought from England and the United States, the expenses being defrayed by subscriptions. During the autumn and winter many hunts were held in the excellent country sourrounding Coblenz.

LIAISON OFFICERS

During the World War each army maintained liaison officers at the headquarters of associated armies. The practice was continued by the armies of occupation, and it was a great factor in the maintenance of mutual understanding and harmonious co-operation.

The A.F.G. maintained at each of the other headquarters a liaison officer who functioned under G-3; the latter was also the ordinary agency through which the foreign liaison offices communicated with the American Headquarters. The liaison officers kept in close touch with the staff of the army to which accredited. They were informed upon all matters of mutual interest, and they were the usual agencies for the communication of routine matters.

The American liaison officers had rare opportunities for the legitimate gaining of information relative to the military affairs and policies of France, Belgium and Great Britian, as well as for the securing information regarding the policies of those countries toward Germany. Colonel Le Vert Coleman, C.A.C., through his initiative and energy and through his friendship with French officers at Mainz, was especially efficient in this regard. He secured a great deal of very valuable military information, including documents and papers containing the ideas of the French army relative to the lessons learned during the World War and French views and policies regarding future organization and training. Using this information he made comprehensive and detailed studies of the latest developments, in practically every arm and branch of the service, that had taken place in the French army, and furnished them to American Headquarters. Copies of these studies, as of all other information secured from liaison officers, were turned over to G-2 who transmitted them to the Military Intelligence Division at Washington.

In addition to their purely military duties, the American liaison officers assisted American travellers, civilians as well as military, in the matter of passport visas, etc.

ROSTER OF OFFICERS

THIRD DIVISION, GENERAL STAFF, A. F. G.

NAME	ASSIGNED	RELIEVED	
MONTGOMERY, Col. John C., G.S.	Nov. 1, 1919	April 9, 1920	G-3, Third Army G-3, A.F.G.
WAINWRIGHT, Major. J. M., Cav.	April 9,1920	Sept. 24,1920	Asst.G-3, Third Army G-3, A.F.G.
LONERGAN, MAJOR T. C., G.S.	Oct. 11,1920	Dec. 20,1920	G-3, A.F.G.
HERR, Major J. K., G.S.	Jan. 27,1921		G-3, A.F.G.
Haislip, Major W. H., Inf.	Nov. 29,1920	July 18,1921	Asst.G-3,A.F.G. Actg.G-3,A.F.G.
COLLINS, Capt. J. L., Inf.	June 15,1920	July11, 1921	Asst.G-3,A.F.G. Actg.G-3,A.F.G.
FECHET, Major d'A., Inf.	Dec. 15,1920	Sept. 1,1921	Asst.G-3,A.F.G. M.G.Officer,A.F.G.
McCARTHY, Capt. L. J., Inf. (DOL)	June 16,1921		Asst.G-3,A.F.G. Rifle & Pistol Mkship. Executive
HODGES, Major C.B., G.S.	June 24,1921	L.A. To U.S.	Asst.G-3,A.F.G.
McCULLOCH, Major W. A., Inf.	Sept. 1,1921		Asst.G-3,A.F.G. M.G.Officer,A.F.G.
DONNER, Major J.W., F.A.	July 23,1921		Asst.G-3,A.F.G.

LIAISON OFFICERS

MIMUTH, Captain H. C., Cav.	Jan. 24,1920 Dec. 2, 1920	Dec. 29,1920	French Army Belgian Army
BARTLETT, Major G.G., Inf.	Feb. 2, 1920		French Army (D.G.C.R.A.)*
BIDDLE, Col. D. H., Cav.	Dec. 31,1919	Jan. 20,1921	British Army
COLEMAN, Col. Lev., C.A.C.	Sept.11,1919 Oct. 31,1919	Oct. 30,1919	British Army French Army
LONGSTREET, Lt. Col. James,Cav.	Jan. 10, 1921		British Army

*Directeur Général des Communications et Ravitaillements aux Armées. (Direct or General of the Communications and Supply of the Aillied Armies--. actually an inter-allied body sitting at Wiesbaden, the main duties of which are indicated by the foregoing title). See Volume II, Chapter 14.

NAME	ASSIGNED	RELIEVED	
WILLIAMSON, Capt. R., Cav.	May 14,1919 Oct. 31, 1919	Oct. 30,1919 Jan. 5, 1921	French Army Belgian Army.

The general supervision of athletic activities was a function of G-3. Becaus of the important part played by the Y.M.C.A, which was under the supervision and control of G-1, there was a close liaison and mutual co-operation between the Athletic Officer (G-3) and the Welfare Officer (G-1). The following officers performed the duties of Athletic Officer:

Lieut. Colonel R. D. Johnson, F.A.
(to July 1919).

Captain A. P. Withers, Inf.
(July 1919 to January 1921).

Captain L. S. Gerow, Inf.
(After January 1921. Also Welfare Officer under G-1).

CHAPTER NO. 20

EDUCATIONAL AND VOCATIONAL TRAINING.

The Army System; Inauguration in the A.F.G.; Operation Results; Officers.

EDUCATIONAL AND VOCATIONAL TRAINING.

THE ARMY SYSTEM.

Prior to 1916 the United States Army had no real system of education of its enlisted personnel. At every post there were schools for the primary instruction of enlisted men, but they were only rudimentary. Usually they were organized and supervised by the Chaplain, the instructors were enlisted men, the attendance was very small, and the results achieved were mediocre at best. There were a few schools for the technical training in certain military specialities of a limited number of men. But not until the passage of the National Defense Act of 1916 was an Army system of general education given serious contemplation.

The entrance of the United States into the World War prevented the system of education from being put into operation. Other than the training of selected men in military specialities, little was done in the way of education of enlisted men.

Shortly after the end of the war steps leading to the establishment of a permanent educational policy were taken, and in September 1919 educational work was placed under the War Plans Division of the General Staff. A broad system of education was outlined in War Department orders, and $2,000,000 were appropriated for the work during the fiscal year 1920. Highly trained educators and technicians were employed to assist in the organization and operation of the system. The purpose of education in the Army was declared to be two-fold -

> To train specialists to meet the Army's needs and to raise the soldier's general intelligence in order to increase his military efficiency.
>
> To fit the soldier for a definite occupation upon

his return to civil life.

A further purpose was to make the soldier a better citizen and a broader-minded man.

The training, except of illiterates, was to be voluntary, but once enrolled in a school the attendance of the soldier became a military duty, and conflict between military and educational training was forbidden by regulations.

INAUGURATION IN AMERICAN FORCES IN GERMANY.

On January 4, 1920, the first schools to operate under the direction of the Educational and Vocational Training Office were inaugurated. These were the Unit Schools, established in all compounds having a sufficiently large personnel to maintain a school. These schools were placed under the immediate control of the post commanders, upon whom devolved the duties of appointing school officers and commissioned and enlisted instructors, of providing suitable and adequate accommodations, of so arranging military duties as to prevent interference with the attendance of students and instructors, and of designating illiterates and "volunteers" to attend school. School officers were assigned the duties of ascertaining the men desiring to attend, securing suitable instructors, organizing the school, arranging classes, and supervising the instruction. The instructors were selected from among the commissioned and enlisted personnel of the post. School supplies and books were drawn from the Educational and Vocational Training Office and from the American Library Association. These schools gave instruction in grammar school subjects. The work of the eight grades of grammar school was divided into three groups, the Elementary, Intermediate and Advanced, and each group was divided into two periods of seven weeks each.

On January 12, 1920, a night school for instruction in the higher academic and commercial subjects was inaugurated in the Kaiser Wilhelm Real Gymnasium (Coblenz), under the direction of the Educational Department of the Y.M.C.A. and supervised by the Educational and Vocational Training Office. Such was the success of this school that it was deemed advisable to establish a central school for higher education. As a result, the General and Commercial School began work in February 1920; inasmuch as funds for the employment of instructors had not been received from the War Department, the Y.M.C.A. assumed the financial obligations. Courses were offered in high school subjects, commercial and Army paper work, and for preparing men for entrance to West Point.

Shortly after the inauguration of the Unit Schools, Service Schools were established. War Department orders required Chiefs of Services to establish and maintain schools of instruction in their various lines of work. The policy was to allow any man, of whatever branch of the service, to receive instruction in these schools either for the purpose of benefiting his lot in the Army or of preparing for a trade or profession in civil life. Schools for training signalmen, company and battery mechanics, cooks and bakers, cobblers, saddlers, horseshoers, draftsmen, surveyors, musicians, chauffeurs, auto mechanics, etc., were established under the direction of the various Chiefs of Services.

In the Spring of 1920 a Mechanical School was organized at Telegraph Barracks (Coblenz) and equipped with machinery which the British had used at a similar school in Siegsburg. In a very short time it was seen that facilities for maintaining a successful school were exceedingly limited at Telegraph Barracks, with the result that the property used by the English Syndicate in Coblenz-Lützel was taken over and a school, equipped to instruct 500 men in the various mechanical trades, was established. Modern machinery was purchased and installed, necessitating an expenditure of approximately $20,000; a large corps of civilian and enlisted instructors was secured, and the real work of the school began in November 1920. The necessity of urgent repairs to motor transportation and the like offered admirable opportunities for instruction and practical work. In the beginning the school was under the direction of the Chief Ordnance Officer, but when Motor Repair Section No. 100 was moved to the Mechanical School compound in the spring of 1921, the school was placed under the Chief Motor Transport Officer.

The instruction of officers in foreign languages was placed under the supervision of the Educational and Vocational Training Office. Every officer (except those specifically excused by the Chief of Staff) was required to take three hours' instruction a week, either in class or in private lessons, during the winter season (November 1 to March 31). During the seasons 1919-20 1920-21 German was compulsory, but in the 1921-22 season officers were given the option of studying either French or German, the great majority choosing French.

OPERATION.

In the beginning comparatively few men applied for courses in the schools. This was primarily due to the fact that the schools were new and their work not fully understood throughout the A.F.G. As a rule organization commanders were not interested, and they made but little effort to induce their men to attend. Even in the case of illiterates, whom organization commanders were required to send to school, the attendance was small. However, by the close of the school year in June 1920 some 1450 men were undergoing instruction in the various schools conducted by the E. & V.T.O.

In order to bring the schools more forcibly to the attention of the men of the command, a personal canvass was conducted in June 1920. All courses were explained in pamphlets and orders, and organization commander were directed to interview personally all men of their organizations, securing from each a statement as to whether or not he wished to attend any of the schools. As a result, 387 men voluntarily applied for entrance to the Unit Schools; 2653 desired to take courses in the Service Schools, and 604 desired higher academic and commercial instruction in the General and Commercial School. This number together with the men already enrolled, totalled 5150, or approximately 32% of the command.

Prior to this time regulations concerning the attendance of men at schools had not been rigidly enforced. With the beginning of the school year in October 1920, orders prescribed that educational and military duties should not conflict and that school hours were to be from 1:00 to 5:00 in the afternoons. Attendance of those enrolled was a military duty, and attention to duty in the schools was just as necessary as on the drill field.

Due to the increased interest, the Unit Schools made great strides toward success. The granting of specialist ratings to instructors, which served to secure a higher type of soldier for this work, and the establishment of a normal short course for instructors, conducted by civilian instructors of the General and Commercial

School, were largely responsible for the improvement of the Unit Schools.

After the summer of 1920 greater interest in the General and Commercial School was manifested. By that time the War Department had allotted sufficient funds to warrant the employment of a civilian faculty of capable instructors, and this force was augmented by those lent by the Y.M.C.A. Approximately 500 men were in attendance at the beginning of the 1920-21 school year, and over $33,000 were expended for this school during that term. For the 1921-22 school year authority for the retention of but one civilian instructor (the former principal) was granted by the War Department, and no more money was allotted. During the summer of 1921 this civilian instructor, with the help of the Educational Department of the Y.M.C.A, conducted a normal course for such enlisted instructors as could be obtained to replace the civilians. These enlisted instructors rendered very satisfactory service during the term beginning in October 1921, the results comparing very favorably with those of the preceding year. Approximately 300 men applied for instruction for the 1921 term, but due to the reduction of the A.F.G. the total number enrolled in the school had been reduced by one-half by December 31, 1921.

The value of the Service School was plainly evident to all by the summer of 1920, with the result that they too had greatly increased enrollments. Their success during 1920-21 resulted in a very gratifying number of applications for the 1921-22 term. In spite of the reduction of the A.F.G. only one Service School, the Band School, was discontinued.

RESULTS.

Prior to December 31, 1921, over 1300 men had been graduated from the various Educational and Vocational Training Schools. These graduates were distributed as follows:

GENERAL AND COMMERCIAL SCHOOLS.................................. 296

(Including stenographers, company clerks, bookkeepers, etc.). In addition, five students taking work in the General and Commercial School succeeded in passing the examination for commission in the army, and five more entered the United States Military Academy. In December 1921 four students were designated to take the final examination for entrance to the Military Academy in 1922.

MECHANICAL SCHOOL..		278
Tire Repairers and Vulcanizers	6	
Auto Mechanics	70	
Motorcycle repairers	1	
Blacksmiths	6	
Chauffeurs	170	
General Machinists	6	
Oxy-acetylene Welders	1	
Mechanical Draftsmen	4	
Auto Ignition	11	
Storage Battery Specialists	3	
SIGNAL CORPS SERVICE SCHOOL................................		129
Telegraph Operators	21	
Telephone Electricians	18	
Linemen	22	
Radio Operators	55	
Power Electricians	13	
ENGINEER CORPS SERVICE SCHOOL..............................		20
Lithographers	2	
Cargador and Packmasters	5	
Topographical and Mechanical Draftsmen	8	
Surveyors	5	
QUARTERMASTER CORPS SERVICE SCHOOL.......................		222
Bakers	110	
Cooks	87	
Wagoners	18	
Cobblers	7	
PROVISIONAL CAVALRY SQUADRON SERVICE SCHOOL................		40
Saddlers	10	
Stable Sergeants	11	
Horseshoers	19	
ORDNANCE DEPARTMENT SERVICE SCHOOL........................		135
Machine Gun Mechanics	64	

Company Mechanics	65	
Battery Mechanics	6	
<u>AGRICULTURAL SCHOOL</u>		11
Dairy Course	3	
General Agriculture	8	
<u>MEDICAL DEPARTMENT SERVICE SCHOOL</u>....................		2
Dental Assistants	1	
Nurses	1	
<u>TYPEWRITER REPAIR SCHOOL</u>..............................		8
Repairing and Overhauling		
<u>UNIT SCHOOLS</u>..		168
	GRAND TOTAL	1309

2. The following list shows the total attendance in the schools as of December 31, 1921:

Post School, Coblenz	171
General and Commercial School	135
Service Schools	143
Mechanical School	214
TOTAL	663

OFFICERS

From the inauguration of Educational and Vocational Training in the A.F.G. until July 1921, Colonel (afterwards Lieutenant Colonel) Frederick S. Young, G.S., was in charge. During the first few months he was assisted by Colonel Charles S. Haight, Cavalry. Other assistants who served under Colonel Young were Captain Emery M. Nourse, Q.M.C., Howard J. Houghland, A.S., Donald L. Dutton, C.A.C., Spencer A. Townsend, Cav., and Marvin C. Keyser, F.A. Major Wentworth H. Moss, Infantry, relieved Lieut. Colonel Young on July 13, 1921, and on the same date Captain Lawrence W. Marshall, Infantry, was detailed as his assistant.

CHAPTER NO. 21

TELEPHONE AND TELEGRAPH COMMUNICATIONS.

Signal Communications during March to Rhine and in the American Area; Antwerp; the Oberpostdiraktion; Cooperation with Allies; Communication with the United States.

TELEPHONE AND TELEGRAPH COMMUNICATIONS.

The World War brought about a great improvement in the American Army in the matter of Signal communications. During the first operations by American troops, the Signal Corps was highly praised for having maintained Signal communications during only two or three hours of a twenty four hour combat; in the later stages of the war the continuous maintenance of communications during a seventy two hour operation aroused no comment, so accustomed to good communication during combat had the A.E.F. become.

Despite the efficiency of the Signal Corps, it was difficult during the march to the Rhine of the Third Army to maintain continuous Signal communications between and among the various units of that command. That Army marched with an extensive front, the various units were widely separated, and the impracticability of establishing a central control caused much confusion and delay. Local circuits were used, but many that apparently were in good condition were found to have been cut or grounded by the retreating enemy.

Upon the arrival of the troops in the Rhinelands and their location in the various army, corps and division areas, the problem became simplified. Maps and charts of the German telephone and telegraph lines were available. Existing circuits were assigned among the units, but as before much confusion was caused by the cutting and "grabbing" of circuits assigned to other organizations.

An immediate censorship was imposed upon all telegraph and telephone communication among the Germans, and all radio communication was absolutely forbidden. All means of communication in any way needed by the American forces were taken over, with the result that many small towns had to get along without any telephone or telegraph communication. Many long distance lines were specialized and, other than for military purposes, all long distance calls were prohibited. The control of Signal communications was exercised through the German organization. In general the German officials co-operated to the best of their ability, but the maintenance of communications still proved a difficult problem and the services of most of the Signal Corps personnel were required for this purpose. Comparatively little trouble was had in radio communication, as the Third Army had its own equipment and the various stations were controlled by the Army control stations.

Until the late summer of 1919 female operators (American) were used in the Coblenz telephone exchange, the code name of which was "Doodlebug". The exchange was known by this name until the early autumn of 1919, when "Coblenz" came into use. Telephonic communication could be quickly had with the following places, among others:

Berlin and other German cities.	Vienna
The Hague	Antwerp
Brussells	Chaumont (G.H.Q.,A.E.F.)
Paris	Tours
London	G.H.Q. French Armies (in France)
G.G.Q. British Armies (In France)	Headquarters of the various Armies of Occupation.

Military telegraph lines led directly to Chaumont, Paris and the various Headquarters in the Rhinelands.

In the autumn of 1919 the direct telephone line to Paris had to be given up (during the summer the American G.H.Q. had moved there from Chaumont), but the telegraph line was maintained throughout 1920 and 1921.

Shortly after the dissolution of the Third Army, the female operators were replaced by Signal Corps personnel and returned to the United States for discharge.

Coincident with the departure of the units of the Third Army, there was a corresponding reduction in the number of specialized telephone and telegraph lines. The policy of the A.F.G. was to retain for military use only those circuits that were actually necessary; the resulting re-establishment of communications between the small towns helped to improve economic conditions in the American area. Censorship was gradually relaxed until it ceased altogether after the Treaty of Versailles was ratified in January 1920. During the period following the signing of the Treaty in June 1919, there had been a gradual return to normal conditions in matters of communications. In the occupied territory the American Headquarters led the way with a view to doing what was best for the economic good of the Rhinelands as long as military interests were safeguarded.

The strict control of the assignment of circuits was maintained, however, and no installations or construction could be made without the approval of the military authorities. All German supplies were kept under close supervision, and in order to preserve them for a possible emergency it was deemed necessary to forbid the installation of even a single telephone without authority from the American Headquarters. After the Inter-Allied Rhineland High Commission came into existence its Ordinances regulated communications matters. These ordinances, as in the case of others, became effective in the American area only after publication in General Orders by the American Commanding General.

Because of the establishment of the Base Port at Antwerp, the American Headquarters secured direct telephone and telegraph communication therewith. In December 1920 the Belgian Government requested the return to Belgian control of so much of the Coblenz-Antwerp circuit as lay within Belgium, and the circuit was released on December 31, 1920. Direct communication by wire ceased, but no great difficulty resulted, as thereafter the necessary communication was maintained by radio.

During the various maneuvers held by the A.F.G., it sometimes became necessary temporarily to suspend the civilian use of particular circuits, in order that they might be used for military purposes. The Oberpostdirektion, the agency of the German Government which administers the telephone and telegraph systems, was quickly responsible to American desires in these cases. No difficulty was experienced, although it became necessary to forbid the various units to tamper with the German lines without specific authority from Headquarters.

In dealings with the Oberpostdirektion the policy was to issue clearly defined instructions and to see that they were thoroughly understood. At the same time every effort to prevent injustice and to provide for the legitimate needs of the Germans was made. The effect was that the Americans secured the co-operation of the German officials, and very rarely was it necessary to caution them relative to the violation of instructions.

The maintenance of communications required frequent conferences with the other forces of occupation. Their needs for lines into and through the American area, American needs for communications with Paris, Antwerp, etc., and the desires of the High Commission, all had to be taken into consideration. The Directeur General des Communications et Ravitaillements aux Armees[1] (at Wiesbaden) was the medium through which the various needs were met and the conflicting interests co-ordinated. In general harmonious relations prevailed in communications matters, as was the case in other questions, and the A.F.G. was well served in this important phase of military administration.

In Coblenz there was maintained a radio station which intercepted despatches from America as well as to the Continent of Europe. Press despatches from Annapolis, Paris, London, Berlin, Rome, Moscow, etc., were received and turned over to G-2, which office published them to the command in the "Radio News" (mimeographed sheets). After the discontinuance, in April 1921, of this G-2 service, the despatches were given to the Amaroc News which published the more important items in its columns.

Official messages between Coblenz and Washington were handled over the Coblenz-Paris circuit, the telegraph office in the American Embassy serving as an agency of transmission between the American Headquarters at Coblenz and the commercial companies in Paris. At one time there was a direct American line between Paris and the American Embassy in London; during that period messages between Coblenz and Washington were sometimes handled over that route. In 1921 the limited funds available caused a change in the method of transmitting messages, and the radio station at Lyons (France) became the relaying medium between Coblenz and Washington. Ordinary messages were handled between Coblenz and Lyons by wire, and between Lyons and United States by radio. Thereafter only secret messages were transmitted by cable.

[1] Volume II, Chapter 14.

OFFICERS

Between December 1918 and August 1919 the position of Chief Signal Officer was held, respectively, by Colonels Alvin C. Voris, Parker Hitt, and John E. Hemphill, all of the Signal Corps. From August 1919 throughout 1921, Major Joseph F. Ware, Signal Corps, was in charge of all Signal Corps activities. His title of Chief Signal Officer was changed to Signal Officer by virtue of the reorganization of the Army in 1920.

CHAPTER NO. 22

SUPPLY AND TRANSPORTATION.

Agencies; Supply - Organization, The General Sales Store, The Quartermaster Farm, Procurement; Food, Fuel, Clothing, Forage and Animals; Transportation - Organization, Embarkation Section, Port of Antwerp; Rail and Inland Water Section - Railway Transportation, Railway Operation, Barge Transportation, Interallied Railway and Waterways Commissions, Miscellaneous; Motor Section - Motor Transport Service; Officers.

SUPPLY AND TRANSPORTATION

AGENCIES

All activities relating to the highly important matters of supply[1] and transportation were supervised and co-ordinated by the Fourth Division of the General Staff (G-4), under which the various Supply Services functioned.

The Quartermaster Corps was charged with the supply of food, clothing, fuel, animals and forage. Until May 1920 the Transportation Branch of G-4 controlled all transportation matters. The Transportation Service was then established, ceasing to be a service in September when it was transferred to the Quartermaster Corps. After that time the latter Service was charged with all transportation activities, which continued to be subject to supervision and co-ordination by G-4.

After the dissolution of the Third Army the disposal of the large stocks of surplus supplies which had been left in the American Zone became an important activity of G-4. The War Department Liquidation Commission established a branch in Coblenz, and G-4 represented the Commanding General in all matters having to do with the disposal of surplus supplies and of enemy war material. The representative of the Liquidation Commission was an Army officer, appointed by the Secretary of War and entitled the General Sales Agent.[2]

Until December 1920 the Base Port at Antwerp operated directly under G-4, when it was placed under the immediate supervision and

[1]This chapter does not deal with pay, ordnance or medical matters.

[2]Volume II, Chapter 11.

control of the Chief of Staff of the A.F.G. In October 1921 it was again placed directly under G-4, and in the same month the Transportation Branch of the Quartermaster's Office was charged with its direct supervision and control.

SUPPLY

Organization

A branch, known as the Supplies Division, of the office of the Chief Quartermaster (later designated The Quartermaster) had as its principal functions the procurement, storage and issue of supplies. The following storage and issue depots were maintained:

Coblenz-Lutzel
Regular supplies, subsistence and clothing

Andernach
Forage

Bendorf
Fresh vegetables, fresh meats, gasoline, oils and greases. The cold storage plant had a capacity of 800 tons.

Supplies were issued from these depots to the railreads at Coblenz, Bendorf, Andernach, Neuwied and Mayen, and subsequently issued therefrom to the troops. During the latter part of 1919 the troops stationed in and near Coblenz were supplied with subsistence stores from the railhead operated under the jurisdiction of the Post Quartermaster, Coblenz. On November 1st, 1920, this system was changed pursuant to verbal orders of the Chief Quartermaster and the issue of subsistence stores to all organizations, as well as all other classes of Quartermaster supplies, was made from the Coblenz-Lutzel Depot.

In the summer 1921 purchase activities, which before had been a function of the Supplies Division, were centered in the Purchase

Branch. Thereafter all purchases were made by the latter office. Excellent results were achieved, duplication was eliminated, as well as the competition among the various branches that had prevailed under the former system. The procedure then followed regulations as closely as the attendant circumstances would permit.

Until October 1921 field bakeries at Coblenz-Lutzel and Bendorf supplied bread for the troops. Marshal ovens having arrived from the United States, a bakery was established at Coblenz-Lutzel, and the field bakeries were discontinued. This bakery supplied the bread for the entire A.F.G., and it proved highly satisfactory as regarded the quality of the bread, economy of production and efficiency in operation.

THE GENERAL SALES STORE. In November 1919 the Coblenz Sales Commissary was moved into the Proviant Magazin, a large and centrally located building formerly used by the German Army. The General Sales Store, as it came to be designated, developed into an actual general store. It comprised a grocery store, a butcher shop, a clothing and tailoring department, and tobacco and candy counters. Many other commodities than those indicated were handled, such as high grade cloth for uniforms, shoes, shirts, underclothes, toilet articles, fresh flowers, etc. etc. Until the summer of 1920 conditions in Germany made it difficult for the American and Allied families to secure food in the German market; the General Sales Store, with the help of the Quartermaster farm, supplied practically all wants; and in general it was of great service to the A.F.G. and the Allied community. The sales privilege was extended to all military personnel and civilian employes of the Allies. During 1920, the period of its greatest turnover, the General Sales Store, with its branches at Andernach and Mayen, supplied innumerable commodities to a community of 24,000 people. Its gross receipts amounted to $200,000 per month handled in German marks, French francs and American dollars.

The Quartermaster Farm. In February 1920 the Educational and Vocational Training Officer[1] established a small farm on a tract of land near Mulheim (north of Coblenz-Lutzel). The original purpose was to use the farm as an agricultural school for the A.F.G. In April the farm was turned over to the Chief Quartermaster who continued it as a school. Many students were enrolled, and courses were given in animal husbandry, dairy husbandry, gardening and general agriculture. The products of the farmers were sold through the General Sales Store. They comprised fresh meats (except beef), poultry, fresh vegetables, eggs, milk and flowers. Rabbits, pigeons and bees were among the specialties of the farm. The dairy proved invaluable, for milk from other sources was scarce as well as of poor quality, and the supply of fresh milk for the sick had been a difficult problem for the Medical Department. The dairy farm supplied sufficient milk for the sick and for American and Allied children, and any excess was sold to adults. The hot beds and flower gardens of the farm enabled the General Sales Store to supply a great many flowers at moderate prices. Altogether the General Sales Store and the farm constituted a feature of the A.F.G. that was not only unique but also added greatly to the general comfort and contentment.

[1]Volume II, Chapter 20.

Facilities for the storage of fuel were established at Bendorf and Andernach, and part of a German coal yard at Coblenz was requisitioned. This yard had excellent rail and water facilities, and was entirely taken over in the late summer of 1921. Thereafter all issues to troops were made from this yard, the more distant organizations being supplied by rail.

In April 1920 the Salvage Depot, which had been located in the forage yard in Coblenz, was moved to Feste (Fort) Franz in Coblenz-Lutzel, where more space was available. At the same time the steam laundry and the shoe repair shops were placed under the direction of the Salvage Officer. Tailor shops and shops for repair work of various kinds were established, and a dry cleaning plant was operated in connection with the laundry. This laundry served practically all the troops of the A.F.G. Through the work of the Salvage Depot a great quantity of property was reclaimed and put back into service, while millions of marks were received for the property sold.

PROCUREMENT

The Third Army had received its supplies from the A.E.F. depots in France, and the large stocks which remained after its dissolution were more than sufficient to meet the immediate needs of the A.F.G. In September 1919 the first request for supplies from the United States was submitted. These supplies did not reach the American Area until January 1920, but the Third Army stocks served to prevent a serious shortage.

Because of the economic conditions that prevailed in Germany during 1919 and part of 1920, the requisition of food, forage and certain other supplies was not resorted to by the A.F.G. Such supplies as did not come from the United States or from Third Army stocks were purchased in France, Belgium, Holland and England. The supplies thus purchased included fish, butter, milk, oats, hay, straw, bran, coal, wood, kerosene, gasoline and motor oils and greases.

As time went on and economic conditions improved certain supplies were secured by requisition upon the German Government. Coal and coke had been furnished by that Government from the beginning (although at one time it was necessary to purchase coal in the United States and England), and in the late summer of 1920 it was called upon to furnish forage also. Potatoes and onions were occasionally requisitions in kind from the German Government, but not until the latter part of 1921 was there a fixed policy in this regard.

The A.F.G. never secured food supplies by direct requisition upon the inhabitants and not until the summer of 1920 was food purchased locally to any extent. When the improvement in economic conditions had reached a point where local purchases of food would not involve hardship to the inhabitants, certain components of the ration were secured in that manner. The necessary funds were requisitioned from the German Government, and the supplies were bought directly from German contractors. These food supplies consisted mainly of fresh vegetables, such as potatoes, onions, beets and cabbage. This practice was begun in August 1920, and was extended to other items as their purchase in Germany became practicable.

In the latter part of 1921 a general change in procurement methods became necessary. The appropriations for that fiscal year (1922) had been subjected to drastic reductions by Congress, and strict ecomony in the expenditure of the funds allotted to the A.F.G. became necessary.[1] It was therefore decided to reduce purchases in Germany so far as possible as to requisition from the German Government such commodities as were obtainable by this method, thus lesseing, to a great extent, the expenditure of United States funds. Purchases made were of those supplies which could not be obtained from the German authorities, or else when there arose an emergency need for small quantities of certain supplies. Under the new policy fresh vegetables (and occasionally other components of the ration[2]), as well as kerosene, gasoline and motor oils and greases, were habitually requisitioned in kine from the German Government.

Food

The supply of food has been discussed already under the heading of procurement. It remains to make brief remark concerning the ration. Until May 1920 the components of the ration had been those prescribed by A.E.F. orders, as modified from time to time to

[1] Funds requisitioned from the German Government had to be deposited to the credit of the United States Treasury, as miscellaneous receipts whence they could be disbursed only in accordance with the Army Appropriation Act of that particular year - that is all disbusements were charged against funds alloted to the A.F.G. by the War Department. Thus the A.F.G. did not differ from other portions of the Army in so far as the enforced necessity for economy was concerned. Volume II, Chapter 9.

[2] Turkeys for Thanksgiving 1921 were secured in this manner and no American funds whatsoever were expended. Those for Christmas 1921 were bought in Germany - this because there were on hand funds that could be used only for the purchase of food. These turkeys cost 13¢ per lb., as compared with 45¢ in the United States.

accord with conditions. Pursuant to the recommendations of a board of officers, the American Headquarters issued an order which established a fixed ration for the A.F.G. The components of this ration were admirably suited to the means of the American troops serving in Europe. These troops were subsisted on "straight" rations and were denied the privilege of making savings from the ration allowance with which to purchase other articles of food. It was not convenient, under the attendant circumstances, for the troops to be subsisted on the authorized garrison ration - hence the "A.F.G. ration" was established. Its principal features were the issue of tobacco (or cigarettes or chewing tobacco) and candy, and increases of 100% in the milk and 33 1/3% in the sugar components.

FUEL

During the latter part of 1919 the coal situation in Germany was serious and the continued supply of the American Area was doubtful. It was deemed advisable to secure coal from other countries. 5000 tons from England and 1500 tons from the United States were received, and arrangements for a continued supply from the United States were made. As the householders of billets occupied by Americans or Allies had to procure their own coal and the supply from German sources was insufficient, they were authorized to purchase American coal - the cost was too great, however, and other arrangements for the heating of billets had to be made. After much negotiation the supply to the A.F.G. of 900 tons of brown coal briquett and 600 tons of coke per month, both from German sources, was assured. Later the situation had improved to such an extent that the monthly supply of 1200 tons of bituminous coal was guaranteed. The arrangements for the shipment of coal from the United States were then cancelled.

While the methods of obtaining fuel for the A.F.G. and public utilities[1] were simple, yet it was difficult to obtain sufficient coal. The German Government had established an agency (Reichskohlenkomission) for the distribution of coal in the occupied territory. Quarterly estimates of the amounts needed were furnished that office. As stocks and labor disturbances, in the coal districts not in the occupied territory, were probable, the policy was to maintain a reserve of from two to three months' supply. To supply current needs and to maintain the reserves, the estimates were closely followed up and the necessary pressure brought to bear. Such were the demands of the Allies for coal, however, and so unsettled were industrial conditions in Germany, that the German authorities had great difficulty in meeting the needs of the A.F.G. and public utilities, to say nothing of industries and householders. The most effective weapon used by the Americans was a threat to requisition, from Rhine barges, coal that was intended for points in unoccupied Germany. The actual seizure of the coal was rarely necessary, as the threat to do so usually brought quick results. Similar methods were used to obtain coal for public utilities, and while the reserves were occasionally reduced to a dangerously low point, they were never exhausted.

The emergency stocks of coal procured from the United States and England were stored at Bendorf. Afterwards it became necessary to maintain a reserve at Andernach also, but the securing of the

[1]Volume II, Chapter 6.

large coal yard at Coblenz-Lutzel[1] solved the problem of storage space for the reserve supply. The storage and issue of coal were greatly facilitated by the use of this yard, and as the cost of transportation was borne by the German Government, there was a considerable saving in the expense. In fact the supply of coal to the A.F.G. was effected without the direct expenditure of any funds whatsoever, and with a minimum amount of clerical work.

In January 1921 the selling of coal from American stocks to German owners of billets occupied by Americans or Allies was discontinued. The German authorities were charged with the supply of coal to these billets, an arrangement that relieved the American authorities of much administrative work and proved generally satisfactory.

Before 1920 wood had been supplied by the German authorities, but in January the supply was interrupted and it became necessary to obtain wood from other sources. Sufficient wood for the winter of 1920-21 was purchased from German contractors. Arrangements were then made for the furnishing of fire wood in kind by the German Government. Thereafter fire wood was secured without the expenditure of American funds.

CLOTHING, FORAGE AND ANIMALS

The procurement and supply of clothing was under practically the same system as in the United States, and therefore will not be discussed.

Prior to April 1920 the main forage yard was at Andernach. This yard had a capacity of 4000 tons and possessed good railroad and

[1]See page

dock facilities. Because of possible floods the yard was moved to Metternich, where adequate facilities were acquired by the laying of additional tracks and the construction of roads. After this hay and straw were stored in the Metternich yard, while all other forage (including a reserve supply of baled hay) was kept at the forage and wood yard in Coblenz.

Animals for the A.F.G. were secured by selection from the large number left by the Third Army, the remainder being sold. Later on a few horses were bought in France and England for special use in polo, etc., but at the end of 1921 the great majority of the animals on hand were those that had come from the Third Army.

TRANSPORTATION

Organization

The transportation activities of the A.F.G. were conducted under the Transportation Branch (G-4) which was divided into three sections. These sections and their respective activities were as follows:

Embarkation:
- Assignment of passengers to transports;
- Preparation of troops for return to the United States;
- Port of Antwerp.

Rail and Inland Water:
- Railway Transportation;
- Railway Operations;
- Barge Transportation;
- Inter-Allied Railway Commission (American Section)
- Inter-Allied Waterways Commission (American Section)

Motor:
- Motor Transport Service.

It will be recalled that the Transportation Branch later became a separate Service, and was finally placed under the

Quartermaster after the control of transportation activities had been transferred to the Quartermaster Corps in 1920. The interior organization and the transportation activities remained practically the same throughout 1920 and 1921, except that the officer in charge had been given the title of Chief Transportation Officer (later changed to Transportation Officer).

Activities

Embarkation Section

The operations of this section were similar to those of the Embarkation Section of the S.O.S. in the A.E.F. Officers, soldiers and other passengers were assigned to transports in accordance with requests from G-1. Arrangements for the transportation of troops from Coblenz, to include arrival on the transport at Antwerp, were made. This included the procurement of the necessary train equipment, the preparation of the troops, etc. Troop movements were made so that the troops arrived at Antwerp on the sailing days of transports, thus avoiding quartering and messing the troops in that city.

Preparation of Troops. Up to the end of 1921 no troop units were returned to the United States, although an average of several hundred individual soldiers sailed on each transport. These soldiers were assembled in the Casual Depot at Coblenz several days, sometimes a week, before the sailing date, and were there organized into provisional detachments. The detachments were called Overseas Casual Detachments and were numbered consecutively. There was one such detachment for each transport, and returning officers were assigned to command it until its arrival in the United States.

The records and personal equipment of each man were inspected before he left the Casual Depot. On July 1, 1921, the Casual Depot was discontinued, after which the inspection of records was performed by the Adjutant, A.F.G. Organization commanders were then made responsible that the equipment of each departing soldier was complete and in serviceable condition.

Port of Antwerp. The base port of the A.F.G. was located at Antwerp, Belgium. Through it passed all replacements and supplies for the A.F.G., as well as officer soldiers, other passengers and freight bound for the United States. The port of Antwerp also served as a point of departure for barges dispatched to Coblenz, this method of transportation being frequently used as a matter of economy. In a few instances freight was shipped from Coblenz to Antwerp by barge, although the usual method of shipment was by rail. A more detailed description of the Port of Antwerp is reserved for a later page.[1]

Rail and Inland Water Section

Railway Transportation. Matters relating to railway transportation were handled through the R.T.O.'s (Railway Transportation Officers). The R.T.O. system had been inherited from the A.E.F., and at the end of 1921 it was still in operation although somewhat reduced in strength and scope. R.T.O's and enlisted assistants were stationed in the railway stations of all important junction points, garrisoned towns and frontier posts, through which much A.F.G. traffic passed. Their main duties were to make all necessary arrangements for the transportation of personnel, equipment and

[1]Volume II, Chapter 23.

supplies, and in occupied Germany to assist the Inter-Allied Railway Commission in the control of the German railways. Thoroughly acquainted with matters relating to railway transportation, the R.T.O's were invaluable in an official sense. They were scarcely less so to the Casual Army traveler, for through them all arrangements were made and journeys were facilitated in every way. Originally only commissioned officers served as R.T.O's, but reductions in personnel caused the officers gradually to be replaced by competent enlisted men, so that by the end of 1921 officers were performing R.T.O. duties at Coblenz, Cologne and Antwerp only. The following list shows the points at which offices for the performance of R.T.O. duties were maintained by the A.F.G.

<u>January 1, 1920</u>

Occupied Germany -	Coblenz, Coblenz-Lutzel, Coblenz-Mosel, Andernach, Mayen, Bendorf, Trier, Cologne.
France -	Paris (Gare de l'Est). Metz.
Luxemburg -	Wasserbillig.
Belgium -	Herbesthal, Brussels, Antwerp

<u>December 31, 1921</u>

Occupied Germany -	Coblenz, Coblenz-Lutzel Coblenz-Mosel, Andernach, Mayen, Bendorf, Engers, Remagen, Trier, Cologne, Aix-la-Chapelle.
France -	Paris (Gare de l'Est).
Luxegbourg -	Wasserbillig.
Belgium -	Brussels, Antwerp.

In France, Luxembourg and Belgium the R.T.O. service was one of liaison with the railways. The offices at Brussels and Antwerp were operated under the supervision of the Port Commander at Antwerp. After the reduction of the A.F.G. had caused the withdrawal of military police from many points, the R.T.O. personnel took over their duties at such places, particularly with regard to the circulation of enlisted men.

The main office of the Railway Transportation section was located at Headquarters in Coblenz, where arrangements for large troop movements and freight transportation were made. Transportation requirements were transmitted to the German railways through the medium of the Inter-Allied Railway Commission. All records relating to transportation accounts were kept in the Coblenz office; these included recommendations as to the amounts to be credited to Germany (as part of the costs of occupation) for transportation furnished the A.F.G.

As there was no through train service between Coblenz and Antwerp and the large amount of official travel warranted it, a daily sleeping car and day coach service between the two points was inaugurated in May 1920. Because of lack of funds this service was discontinued on June 30, 1921, and thereafter was operated only in connection with the arrival and departure of transports.

During the World War and for some time after the Armistice the customs barriers on the frontiers of the Allied countries had been raised in so far as the Armies were concerned. After the ratification of the Treaty of Versailles the barriers were again put down, but arrangements were made to exempt members of the A.F.G. and their legitimate possessions from the payment of customs dues. From time to time civilian tourists would endeavor to secure under false pretenses, the privileges extended to the A.F.G., or, soldiers would attempt to smuggle prohibited articles, all of which caused embarrassment to the American authorities.

There was no expense incident to railway transportation in occupied Germany, as the German Government furnished it under the terms of the Armistice Agreement and the Treaty of Berlin (November 1921). In France and Belgium, of course, transportation expenses had to be paid. Under the Parker-Delacroix Agreement of November 7, 1919, Belgium agreed that the A.F.G. were to be on the same basis as the Belgian Army with regard to rates and privileges on the Belgian railways. When the Graves Registration Service began operations in Europe, it was placed on the same basis. As the Agreement covered all transportation for the American Army, including military attaches and welfare organizations, the American Forces in France continued to issue transportation warrants for travel on Belgian railways. While careful records were kept by the A.F.G. and the Graves Registration Service, an accurate audit of all the accounts was impracticable. Therefore, the Belgian Government was requested to furnish a statement, based on the records of the Ministry of Transportation. That Government admitted that it did not have the necessary facilities and requested that the American Army make a "bulk settlement". The A.F.G. declined to make such a settlement, but offered to place American auditors in the Ministry of Transportation for the purpose of compiling the accounts. This was agreed to; the work was begun in March 1921 and completed in July. The audit was based entirely on the records in the Ministry of Transportation, supplemented by the records of the A.F.G. On July 30, 1921, a settlement was made of the accounts covering the period from November 1919 to June 30, 1921.[1] Rates satisfactory to the A.F.G. were secured in the conference preceding the settlement; these were incorporated therein and were to remain in force until the expiration of the Parker-Delacroix Agreement on November 7, 1922. After July 1921 the accounts for transportation on the Belgian railways were prepared monthly by an American auditor from the records of the Belgian Ministry of Transportation.

Railway Operation.[1] The presence in the American area of railway operating troops materially affected the attitude of the German railway employees towards strikes ordered by labor leaders in unoccupied Germany; the knowledge that in case of a strike the Americans could operate the railways sufficiently to maintain their communications seemed to have a salutary effect. Up to December 31, 1921, no strike of sufficient extent to make necessary the intervention of railway operating troops had occurred. These troops never came under the control of the Transportation Service, as they would have done had they been called upon to operate as railway troops. Their administration and training was supervised by the Engineer Officer, A.F.G., although the Transportation Service cooperated in the matter of securing railway facilities for their technical training.

[1]The welfare organizations paid their portion of the account through the American Headquarters.

[1]Volume II, Chapter 25.

Barge Transportation. Up to July 1921, as a measure of economy, the greater portion of the supplies from the United States were shipped by barge from Antwerp to Coblenz. During that period perishable supplies and those that were required quickly were the only kinds shipped by rail. The barge shipments were made under contract with the Belgian Government. The contracts provided that the necessary stevedoring in Antwerp and the American area should be provided by the A.F.G., and that the responsibility for the transportation after reaching German territory should rest upon the A.F.G. The Transportation Service handled all barge shipments, including the delivering of the supplies to the American depots, until April 1921 when the duty of receiving, checking and delivering supplies in the American area was transferred to the local Quartermasters. German barges were used only for the transportation of coal and a few other supplies originating in German territory. The low rates offered by the Belgian Government, and its desire to have in Belgium as little German transportation as possible, made the use of German barges in Belgium inadvisable as well as unnecessary. After July 1, 1921 because of the low railroad rates offered, only a few shipments were made by barge.

Inter-Allied Railway and Waterways Commissions.[1] The activities of the Transportation Service in connection with these Commissions have been discussed in an earlier Chapter.[2] In order to reduce the overhead, in September 1920 the R.T.O. service was combined with the American Section of the Inter-Allied Railway Commission; the Assistant (in charge of rail transportation) to the Chief Transportation

[1] A.M.G., Chapter 3, Volume I, (page 59).

[2] Volume II, Chapter 14.

Officer was then made President of the Franco-American Delegation of that Commission. (This Delegation controlled the railways in the American area.)

Miscellaneous. The steamers used for sight-seeing trips on the Rhine, under the auspices of the Y.M.C.A., were operated under the supervision of the Transportation Service.

When the huge amount of railway transportation equipment was sold to France by the A.E.F. in 1919, Hospital Train No. 70 was withheld from the contract and turned over to the A.F.G. This train, together with 25 refrigerator and 10 tank cars which had been rented from France, were in use by the A.F.G. throughout 1920 and 1921.

Two unusual activities were the operation of special trains to the battlefields in 1920, and the dispatching of a train to Leipzig for the relief of Americans and British who had been stranded there during the disturbances following the Kapp Putsch[3] in March 1920.

Motor Section

Motor Transport Service. Upon the departure of the divisions and other units of the Third Army in the spring and summer of 1919, their motor transportation was turned over to reception parks. After the A.F.G. had been thoroughly equipped, the surplus transportation was sold to Motor Organizations, Limited, a British firm.

After the arrival of the Provisional Infantry Brigade in November 1919, six motor transport companies and three service

[3]Volume II, Chapter 1.

park units (two of which were provisional) were available for the operation of the Motor Transport Service. A mechanical school and a chauffeurs' school were conducted under the supervision of the Motor Transport Officer. By the end of 1921, in conjunction with the general reduction of the A.F.G., the Motor Transport Service had been reduced to two motor transport companies and one repair shop. The reductions caused the employment of an ever increasing number of German civilians, until by December 31, 1921, practically all repair work was being performed by them. Up to August 1921 these employees had been paid from American funds, but after that time, as requisitioned labor, they were paid directly by the German Government.

In addition to serving the A.F.G. the Motor Transport Service did much repair work for the Graves Registration Service, American military attaches, American embassies and legations, and various American relief organizations. (Because of the favorable rate of exchange, many spare parts could be purchased in local market at comparatively low prices.)

There was a gradual reduction in the amount of motor transportation throughout 1920 and 1921, especially in the latter year when there was a reduction of approximately 50%. 36 vehicles were sold during 1921, while in the early autumn of that year 79 Cadillac touring cars and limousines were returned to the United States. The vehicles on hand and in operation at the end of the calendar years 1920 and 1921 were as follows:

Year	Passenger Cars		Other Vehicles	
	On Hand	Operating	On Hand	Operating
1920	152	103	800	678
1921	99	72	678	271

OFFICERS

From the formation of the A.F.G. until September 1920 Major (formerly Lieutenant Colonel) Brehon Somervell, G.S., served as A.C. of S., G-4. Durinng the interim of two months between the departure of Major Somervell and the arrival of Major Herr, Major D. A. Robinson, Cavalry, acted as G-4. He was relieved by Major John K. Herr, G.S., who served as G-4 until January 1921, when he became G-3. Lieutenant Colonel H.H. Sheen, Q.M.C., in addition to his duties as Quartermaster, was appointed G-4, to succeed Major Herr, in which capacity he was still serving at the end of that year. During 1920 and 1921 Majors R.L. Miller, A.S.C., E.C. McGuire, Cavalry, and R.B.Patterson, Cavalry, in the order named served as Assistant G-4.

Supply

The Supply Branch of G-4 was disbanded in November 1920. Up to that time Major A. R. Harris, G.S., and Captains Joe N. Dalton, Infantry, and W.C. McMahon, Infantry, had served in that branch.

Colonel George Mc.K. Williamson, Q.M.C., was Chief Quartermaster of the A.F.G. until early September 1919, when he was relieved by Lieutenant Colonel (than Colonel) H. H. Sheen, Q.M.C., who continued in that capacity throughout 1920 and 1921. The names of those assistants who were more prominently connected with supply activities, are set forth below:

Lieut. Col. R.E. Grinstead, Inf.		Executive
" " C.C. Reynolds, C.A.C.		Purchases

Page 15 - It is suggested that the following named commissioned officers be included in group shown as functioning under the Chief Quartermaster:

Major R. Talbot, Jr.	Depot Q.M.
" E. Montgomery.	Salvage
Captain H.F. Gardner	Property
" J.W. McDonald, Cav.	"
" E.W. Austin, F.A.	Coal
Captain T.J. McGrath, Q.M.C.	Executive
" E.M. Scott, Q.M.C.	Purchases
" E.E. Hagen, Q.M.C.	Subsistence
" R.F. Smith, Q.M.C.	"
" M.A. Lowenberg, Q.M.C.	General Sales Store
" H. Feldman, F.A.	Labor
" A.B. Proctor, Q.M.C.	"
" H. Haney, Q.M.C.	Salvage

Transportation

The Transportation Branch, G-4, was headed, respectively, during 1920 by Majors J. B. Crawford, C.A.C., and Abbott Boone, Cavalry. Captain W.H. Sadler, Q.M.C. was in charge of rail transportation and Colonel R. L. Roach, I_fantry, was Chief Motor Trsns-Officer.

In September 1920 when transportation activities were placed under the Quartermaster, Major W. A. McCain, A.M.C., became Transportation Officer and continued as such until his return to the United States on leave in February 1922.

The following officers performed R.T.O. duties during portions of the years 1920 and 1921:

Maj r C.M. Fra ster, Inf.
" W.J. Wrona, F.A.
Captain M.K. Barroll, Jr., Q.M.C.
" R.L. Creed, Cav.
" T.C. McCormick, F.A.
" H.C. Minuth, Cav.
" O.W. Reed, Inf.
" W.B. Van Auken, Cav.

1st Lieut. E.G. Bomar, C.A.C.
" " M.S. Courtney, T.C.
" " L.M. Kreber, F.A.
" " N.McG. Maner, Inf.
" " L.J. Whitlock, F.A.

Major H. J. Weeks, Q.M.C., was appointed Motor Transport Officer in September 1920, and served in that capacity throughout 1921.

CHAPTER No. 23

THE PORT OF ANTWERP.

Establishment; Organization; Water Transportation; Rail Transportation; Miscellaneous; Relations with the Belgians; Officers.

THE PORT OF ANTWERP

Early in 1919 it was planned by G-4 of the A.E.F. to use Rotterdam and the Rhine river for the supply and evacuation of the units of the American Army of Occupation. The permission of the Dutch Government was secured, but the activities at Rotterdam were confined largely to the movement of supplies, including the shipment of Ordnance materiel to the United States. In March 1919 the base port was enlarged to include Antwerp, and was given the title of Base Section No. 9, becoming generally known as the Antwerp-Rotterdam Base.

The plan to use Base Section No. 9 and the Rhine for evacuating the units of the Third Army was abandoned, and all troops were sent by rail to Brest and other American base ports in France, whence they sailed for the United States. Rotterdam was abandoned during the summer of 1919 and Brest ceased its activities as an American base in December 1919. Antwerp then became the only American base in Europe and was called the Port of Antwerp. Its organization was generally similar to that which had been used at base ports in France. From the beginning until December 1919 its main activities were the receipt and transmission of supplies for the Third Army and the A.F.G., the evacuation of certain supplies to the United States, and the storage of supplies[1] that had been accumulated for the use of the Third Army. Shortly thereafter Antwerp [illegible] the only European port of call for Army transports, and with that it began to function in every way as a base port.

[1] These supplies became surplus and were sold by the War Department Liquidation Commission.

ORGANIZATION

In charge of the Port of Antwerp was a field officer known as the Port Commander. Under him there were the following sub-divisions:

Headquarters Commandant
Army Transport Service
Quartermaster Corps
Rail Transportation and Baggage
Embarkation and Landing
Rents, Requisitions and Claims
Port Statistician
Port Inspector
Medical Department
Assistant Provost Marshal
Signal Corps
Postal Express Service.

Local conditions prevented the concentration of the various sub-divisions of the Port of Antwerp. During 1920 and 1921 more shipping arrived at and departed from Antwerp than ever before, there had been a large influx of people from the Belgian devastated districts, and the facilities could not meet the increased demand. Headquarters were located in a building which was sequestered German property and which was rented from the Belgian Government at a reasonable rate. A dock was rented from the city of Antwerp, but it was some two miles from the Headquarters Building. A reservation near the dock was rented, and on it were built adrian barracks and other temporary structures for use as dining rooms, bath rooms, amusement hall, guard house and garage. When it became necessary to establish a hospital, the only suitable building that could be secured was two miles from the Headquarters Building and four miles from the dock.

WATER TRANSPORTATION

The water transportation activities were of two classes, those dealing with seagoing vessels and those having to do with non-seagoing craft.

The seagoing vessels were principally Army transports, while space on commercial vessels was used from time to time for the shipment of supplies. Up to the end of 1921 fifty four Army transports had called at the Port of Antwerp, and fifteen shipments by commercial vessels had been received. 61,885 kilo-tons of cargo had been received and 15,000 tons had been shipped to the United States. 23,037 passengers had sailed by Army transports and 9,650 had arrived. The bodies of 28,000 American soldiers had been shipped from Antwerp and 12,000 from other European ports. The operating costs of the 54 Army transports were $469,660.66; the expenditures on account of Army transports at other European ports[1] during the period (1920-21) amounted to $189,000.

For a short time after the organization of the Base Port considerable difficulty was caused by the rivalry of British, Dutch and Belgian pilots. If a transport employed a British pilot on one trip and a Dutch pilot on the next, jealousy would be aroused, and the result would be a partial boycott of American transports. With the assistance of the Belgian Ministry of Marine, an arrangement was made with the Belgian Pilots' Association whereby only experienced Belgian pilots were employed, at rates stated in the agreement. After that no further difficulty was experienced.

In order to prevent interference with Army personnel passing through the port, the question of Customs duties had to be settled.

1 These costs were paid by the Port of Antwerp.

Arrangements were made with the Belgian Ministry of the Interior to permit American military personnel and supplies to pass in and out of Belgium without the usual delay. The Collector of the Port was furnished a certificate, enumerating the quantity and kind of cargo and its destination, and the number of passengers and their destination; transports could then enter and clear without molestation.

Much of the cargo shipped from Antwerp to Coblenz was sent by barge.[1] Barges were towed down the Scheldt river to Hansweert and thence by canal past Rotterdam and Emmerich to the Rhine. Part of this voyage was through Holland.

Barging was done under a special contract with the Belgian Ministry of Transportation. In order to permit American cargo to pass through Holland without delay, the permission of the Dutch Government for American shipments to be transported in bond was secured. The Dutch customs authorities were prohibited from opening any barges carrying American cargo except in the presence of an American officer. The Belgian Government then accepted responsibility from the time the hatches were sealed at Antwerp (or at Coblenz, as the case might be) until the arrival of a barge at its destination; if the seals remained intact until arrival, the Belgian Government was released from responsibility.

There were several instances of the robbery of barges by "river pirates" on the Rhine; this caused guards to be placed on barges during their voyage in German waters.

[1] Volume II, Chapters 14 and 22.

During 1920 and 1921 there were 91 barges despatched from Antwerp to Coblenz; these carried 37,211 kilo tons of cargo. 7 barges, carrying 1,755 kilo tons were sent from Coblenz to Antwerp. The total cost was $146,070.20. During the same period three barges were used at Antwerp for lighterage purposes; the total cost was $278.05.

RAIL TRANSPORTATION[1]

When the Port of Antwerp began to perform all the functions of a base port the American Forces in France were still in existence, the Graves Registration Service was beginning its operations, and the A.F.G. were 25,000 strong. Shipment by rail to and from all parts of Europe was made, requiring a definite agreement with the Belgian Government regarding railway transportation. This became known as the Parker-Delacroix agreement.[2] Under it the French system of Ordres de Transport (or O.D.T.'s), which correspond to the American Transportation requests, was adopted, and it continued to be used throughout 1920 and 1921. The O.D.T.'s were issued to individuals or organizations by the American authorities for so much of the travel as was in Belgium. Copies were kept by the American authorities; as each O.D.T. was given a consecutive number, duplication was prevented. Periodical audits were made by the Port of Antwerp, these audits forming the basis of the payments to the Belgian Government. After June 30, 1920 no direct payments were made, as the Belgian Treasury Department had requested that the amounts owing to Belgium should be set off against the Belgian debt to the United States. Negotiations to

1 Volume II, Chapters 14 and 22.
2 Volume II, Chapter 22.

that effect were instituted, but they had not been entirely completed by December 31, 1921.

During the 26 months ending December 31, 1921, the following had been transported over Belgian railways, at a total cost of $160,300:

- 18,000 troops and other personnel
- 28,000 bodies
- 27,000 caskets
- 1,500 tons of freight
- 1,000 kilo tons of baggage
- 5,000 (approximately) of the Provisional Infantry Brigade (through Belgium, en route from Brest to the Coblenz bridgehead).

MISCELLANEOUS

The Port of Antwerp had much to do with the repatriation of Polish-Americans from Danzig and of Czecho-Slovaks from Siberia to Trieste. Representatives and personnel were sent to both ports, and the accounts were settled by the Antwerp base port. Eight Army transports called at Danzig and carried 12,018 Polish-Americans to the United States, at a cost of $148,004. Twelve steamers carried 36,500 Czecho-Slovaks to Trieste and brought in 10,000 tons of cargo; the total cost was $901,357.

Representatives of the Graves Registration Service were stationed at other European ports of call for the transports carrying the bodies of American soldiers. They transmitted all bills to Antwerp whence the accounts were paid.

In June 1920, at the request of the Graves Registration Service, the Port of Antwerp chartered a steamer which was sent to the islands of Mull and Muck, north of Scotland. On these islands had been buried the bodies of some 500 American soldiers and sailors who had perished when the "Otranto" was torpedoed by a German submarine during the World War. The bodies were disinterred and shipped to Liverpool, whence they were carried to the United States on an Army transport.

In general the supplies for the Port of Antwerp were secured from Coblenz. Ocassionally fresh vegetables were purchased in the local market. Supplies for transports were obtained from stocks on hand at the Base Port or by purchase in Antwerp. Coal for the Port was generally obtained from Coblenz, but coal and fuel oil for transports were purchased locally. During 1920 and 1921 purchases, amounting to $1,253,000, of potatoes, onions, butter, fresh fish, gasoline and various exceptional articles for the A.F.G. were made.

RELATIONS WITH THE BELGIANS

The Antwerp officials and those of the Belgian Government were most considerate in their dealings with the American authorities of the Port of Antwerp, and they were of great assistance in many ways. The most cordial relations prevailed, as was also the case with regard to the personnel of the Base Port and the population of Antwerp.

O F F I C E R S

Major (then Lieutenant-Colonel) W. C. Koenig, C.A.C., was placed in charge of the Port of Antwerp in November 1919. In accordance with orders from Coblenz, he assumed the title of Port Commander on December 1, 1919, remaining on that duty throughout

1920 and 1921.

The following officers, among others, were on duty at the Port of Antwerp during 1920 and 1921:

Lieut. Col. F. C. Johnson, Q.M.C.
Major Abbott Boone, C.A.C.
" W. A. McCain, Q.M.C.
" R. P. Hall, C.A.C.
" G. R. Meyer, C.A.C.
" W. H. Moss, Inf.
" J. L. Parkinson, I.G.D.
" F. E. Tibbetts, F.A.
Captain L. C. Dill, Cav.
" H. V. Hand, Inf.
" C. H. Morgan, C.A.C.
" R. Orsinger, Inf.
" A. T. Rich, Inf.
" P. L. Singer, Cav.
" J. C. Waddell, C.A.C.

CHAPTER NO. 24

MILITARY CONSTRUCTION.

Duties; Personnel; Labor; Material; Work performed; Costs; Officers.

MILITARY CONSTRUCTION

The Army Regulations assign the following duties, among others, to the Corps of Engineers:

> In time of actual or threatened hostilities, within the theatre of operations, it has charge of the location, design, and construction of wharves, piers, landings, storehouses, hospitals, and other structures of general interest; and of the construction, maintenance, and repair of roads, ferries, bridges, and incidental structures; and of the construction, maintenance, and operation of railroads under military control, including the construction and operation of armored trains.

In compliance with this paragraph the Engineer Office, A.F.G., was charged with all military construction and with the repair and maintenance of existing military buildings and posts.

The permanent personnel of the Construction Section of the Engineer Office consisted of one Lieutenant (detached from Company A, 1st Engineers) and six Master Sergeants. Dependent upon the amount of work to be done, enlisted men of the various grades were detailed from the Engineer units.

The Construction Officer secured the necessary labor from three sources--

> Engineer troops;
>
> Hired German labor, supervised by Engineer soldiers;
>
> The Reichsvermogensamt (an agency of the German Government, charged with the furnishing and upkeep of quarters used by the forces of occupation), under the supervision of Engineer personnel.

When Engineer troops were used, both materials and tools were furnished by the Engineer Depot, while only the material was furnished when work was performed by hired labor. When construction was performed under contract, payment for both material and labor was made directly by the Reichsvermogensamt.

A large amount of construction and many repairs were necessary at the outset, as the installations in the German barracks had to be adapted to American organization and needs. Battalion kitchens without mess halls were found, while some barracks had battalion baths and others had none. Only the more modern barracks had wash rooms, and in many cases the toilet facilities were crude. The interiors of the majority of the barracks had been left in very bad shape by the Third Army. A complete renovation of all buildings was required to make them habitable, and in many cases new barracks and mess halls were constructed. Finally, the stable facilities did not correspond to the American organization.

The principal construction work performed under the supervision of the Engineer Office was as follows:

> Alteration and enlargement of the hospital and officers quarters at Antwerp, and the construction there of a garage and barracks.
>
> Construction at Weissenthurm Flying Field of a small post, with water sewerage and lighting systems. The buildings included barracks, officers quarters, headquarters, hangars, laboratories and a garage.
>
> Barracks for one company at Kreusberg.
>
> Barracks for a headquarters platoon and one company (16th Engineers) at Coblenz.
>
> Mess Halls and stables in 2nd Brigade area.
>
> Stables for Infantry machine gun companies.

After July 1920 almost all construction was done under contract and paid for by the Reichsvermogensamt; and on April 30, 1921, all construction duties were transferred to the Quartermaster Corps.

The construction, repair and maintenance of barracks and quarters for the A. F. G. entailed practically no expense to the United States. The troops were quartered in barracks owned by the German Government or in buildings that were requisitioned by that Government and placed at the disposal of the A. F. G. In addition to performing the construction ordered by the American authorities, the Reichsvermogensamt kept in repair all barracks and buildings used by the A. F. G., and defrayed the expenses of maintenance.

It may be said in conclusion that the European system of quartering troops in barracks located in cities, as compared with the isolated Army posts used in the United States, is both economic and convenient. The adoption of such a system in the United States would probably result in a large reduction in annual military expenses, as the initial expense of construction could be defrayed by the sale of the military reservations that could be abandoned.

OFFICERS

During the period before military construction activities were transferred to the Quartermaster Corps, 1st Lieutenants Frederich F. Fresh and Clarence N. Iry, and Master Sergeant Joseph A. Hoeft (all of the Corps of Engineers), respectively, were in charge of the Construction Section. Afterwards Major Clifford L. Corbin and Captain George F. Hobson (both of the Quartermaster Corps) were charged with those activities.

CHAPTER NO. 25

MILITARY RAILWAY OPERATION.

Railway Units; Training and Operation; Plans; Officers.

MILITARY RAILWAY OPERATION

RAILWAY UNITS.

It is obvious that a military force in occupation of foreign territory should be able to operate the railways sufficiently, in case of need, to insure its communications and to meet its ordinary needs. As the original organization of the American Forces in Germany had not provided for railway troops, the War Department was requested in the early autumn of 1919 to assign a detachment of such troops to these forces. This request was granted, and Company B, 1st Engineers (then stationed in Virginia) was reorganized as a standard gauge railway operating unit. Its authorized strength was 6 officers and 253 enlisted men, "for possible railway operations in France and Germany," and it was to be composed of experienced railroad men. Approximately 75% of the men enlisted for one year, the remainder for three years.

The company reached the American area on December 30, 1919. It was prescribed that its administration and training would be under the Chief Engineer (later designated as Engineer Officer), while its railway operations would be controlled by the Assistant Chief of Staff for Supply (G-4). Authority to attach to the Company all qualified railroad men in other units of the A. F. G. was granted, and the Trasnportation Service Detachment (50 men) was assigned to it for a period of seven months.

In late 1920, due to the expiration of the terms of enlistment of the large majority of its personnel, the early disbandment of the company as a railway unit became necessary. The remainder of its personnel were transferred to the 16th Railway Engineer Battalion (later designated as the 16th Engineer Battalion), which came into existence on January 27, 1921, and the company itself was reorganized and recruited as a sapper unit.

The authorized organization of the 16th Railway Engineer Battalion provided for a headquarters and supply platoon and three companies, but there were sufficient officers and men for only two companies. The maximum actual strength of the battalion at any time was 250 men, distributed among the Headquarters and Supply Platoon and Companies "A" and "B." To accord with local conditions, facilitate training and cause each unit to be self-sustaining, each company was organized as a combined shop and operating unit.

The reduction of the United States Army caused the battalion, with the exception of Company "A," to be placed on the inactive list on November 1, 1921. A strength of 3 officers and 90 men was authorized for Company "A," which was then reorganized with a view to the taking over, as far as possible, of the duties theretofore performed by the battalion.

TRAINING AND OPERATION.

In conjunction with their technical training, the railway units of the A. F. G. underwent a certain amount of military training, including infantry drill, interior guard duty, calisthenics, lectures on military subjects, and pistol practice.

It was evident that without delay the personnel of the railway units should become familiar with the German railways and their methods of operation, and that they should secure a good knowledge of all railway equipment and facilities in the American area and at the important stations in the adjacent French and British zones. To this end the Coblenz Delegation (which had an American officer as President) of the Inter-Allied Railway Commission acted as an intermediary between the American railway units and the German railway authorities, furnishing to the units all necessary maps, plans, rules, regulations, etc. In the beginning the President of the Coblenz Delegation was more or less independent as regards control by the American Headquarters. Later he was placed under G-4, a change which greatly facilitated the training of the railway units.

The technical training of the railway troops comprised the following:

> Observation of the German railways and their methods of operation.
> Study of station and yard facilities in the American zone and at important points in adjacent occupied areas.
> Procurement of data for use in case the military operation of the railways became necessary.
> Operation of switch engines at Andernach and Bendorf.
> Operation of school trains.
> Schools of instruction.

Observation parties, conducted by technical railroad men, made tours of the stations, yards, roundhouses and machine shops at the large stations in the occupied zones, and in this manner acquired a good general knowledge of the railway system in the areas occupied by the American, French, and British troops. Enginemen rode as observers on German locomotives, covering the entire occupied territory in this way. Signal men were stationed in signal towers and received practical instruction in German signal methods. In general, through technical instruction and personal observation and experience the railway troops were made familiar with every phase of German railroads in the occupied territory. At the same time, advantage of the opportunities afforded was taken and complete data of the railroads in the American zone were secured and recorded.

At Andernach were stationed the Headquarters of the 2nd Brigade, A. F. G., and Headquarters and two battalions of the 5th Infantry, and at Bendorf was a large Quartermaster depot. At both places the operation of switch engines by the railway units afforded excellent opportunities for the training of enginemen and yard crews in yard work. On December 31, 1921 the switch engine at Bendorf had been continuously operated by American personnel for nearly two years.

For practical training it was decided to procure facilities for the actual operation of trains by American railway personnel. It was impossible under the Armistice Agreement or the Treaty of Versailles to requisition either a branch railroad or its equipment for use in the training of military railway units, but upon request the German railway authorities courteously placed the necessary facilities at the disposal of the Americans. It was agreed that the expenses involved should be deducted from the costs of the American occupation. The expenses included the salary of the German "pilot," a flat rate for locomotives and cars and the value of the fuel, oil and roundhouse supplies expended. The bills rendered by the railway authorities were checked and approved by the Americans before being charged against the costs of occupation. The agreement was put into writing, and it was arranged that in all minor matters there should be direct dealings between the railway troops and the German authorities. Important matters were taken up with the Germans through the Engineer Officer and the President of the Coblenz Delegation, I.A.R.C.[1] The trains operated under this agreement were called "American School Trains." On each was a German "pilot," who was an experienced locomotive engineer; he represented the railway authorities, acted in an advisory capacity observed the work of the American train crew, and upon request gave them the benefit of his experience. The train crews were composed entirely of Americans, but station and signal personnel were German. Before operating on a school train, the American soldier was thoroughly examined as to his technical knowledge by an officer of the American railway troops and a German railway official.

School trains were run on two branch railroads in the American area. A company (later a platoon) of railway troops was stationed at Kreuzberg in the Ahr Valley; between that town and Junkerath, a distance of 35 miles, three trains were operated daily except Saturdays, Sundays, and holidays. These were practically dummy trains, as except in case of emergency they carried no freight or passengers other than railway employees and American railway personnel. The other company (later reduced to a platoon) was stationed at Kottenheim; it operated three trains daily except Saturdays, Sundays and holidays, between Andernach and Mayen, a distance of 15 miles. These trains carried members of the A.F. G. and associated armies and the personnel of the welfare organizations and the Inter-Allied Rhineland High Commission; as troops were stationed at both places, a comparatively large number of passengers was handled.

The operation of these trains followed the methods used in

[1] Volume II, Chapter 14.

the United States on a railway division, and all personnel engaged in the work were given appropriate railroad titles. The necessary repairs to locomotives and rolling stock were made by the American personnel, the roundhouse, shop and yard facilities at Kreuzberg and Mayen being used for this purpose.

The schools of instruction were conducted by experienced railroad men, and the students were prepared to act as conductors, brakemen, engineers, and firemen. There was a complete course in German railway operation, and instruction in American railway practice was given as well.

While in itself the railway training involved railway operations, the American railway units engaged in many other activities. Rail and water shipments between the Base Port (Antwerp) and Coblenz were convoyed, as were shipments from Coblenz to various European cities where American embassies or legations, commissions and relief organizations were located. Cars were convoyed to and from German coal districts. Every troop train between Coblenz and Antwerp was provided with American conductors, mechanical observers and electricians, and military coaches on regular trains were furnished with conductors and guards. The railway units provided the necessary personnel at points where R.T.O's (Railway Transportation Officers) were stationed, and furnished conductors and guards for the official car of the Commanding General, and for Hospital Train 70. The hospital cars were stationed at Kreuzberg and Mayen, where they were kept in repair by American railway personnel. The railway units also furnished men for inspecting locomotives and railway equipment in general, for the I.A.R.C.

PLANS.

In the event that a strike or the outbreak of hostilities or grave public disorder should make necessary the actual military operation of railways in the American zone, the American railway troops were to function under the Chief Transportation Officer as the representative of G-4. Agreements with the French as to the lines and trains that in such a contingency should be operated were drawn up; these plans were modified from time to time as the strength of the American railway personnel was reduced. (A railroad strike in the Saar District having caused the French to operate trains there, American personnel were sent to observe the French methods.)

In December 1921 a large number of the German roundhouse, shop and yard employees at Coblenz walked out. American railway personnel was immediately sent to the three Coblenz stations and other important yards in the American area, to insure the operation of trains required by the military authorities and to prevent sabotage of railway equipment. There arose no necessity for further intervention by the Americans, as conditions returned to normal within a very few days.

OFFICERS.

The following officers of the Corps of Engineers were engaged in military railway operations during 1920 and 1921:

Major Edwin A. Bethel,
*Captain J. E. Brown,
Captain Harry O. Tunis,
Captain Walter D. Luplow,
Captain Cecil R. Moore,
Captain Robert A. Radford,

*Discharged or transferred to another arm.

 1st Lieut. John M. Harmon,
 1st Lieut. Chester C. Hough,
 1st Lieut. Clarence N. Iry,
*1st Lieut. French W. Lake,
 1st Lieut. Maybin H. Wilson,
*1st Lieut. Ralph Woolsey,
*2d Lieut. Henry A. Haenseler,
*2d Lieut. James C. Rickner.

*Discharged or transferred to another arm.

CHAPTER NO. 26

THE MEDICAL DEPARTMENT, A.F.G.

Organization; Polish Typhus Relief Expedition; Hospitalization - Station Hospital, Convalescent Hospital; Sanitation - Water Supply, Food Supply, Disposal of Waste, Insects; Diseases - Measles, Influenza, Pneumonia, Venereal Disease; Office of the Attending Surgeon; Medical Supplies; Field Units; Dental Service; Veterinary Service; Officers; Appendix X - Venereal Disease in Coblenz in 1811.

THE MEDICAL DEPARTMENT, A.F.G.

ORGANIZATION.

After the departure of the Third Army and the formation of the "permanent" occupying force which was designated as the American Forces in Germany, the units of the Medical Department had been correspondingly reduced until its final organization, after the arrival of the "Silesian Brigade" (afterwards designated as the 2nd Brigade, A.F.G.), was as follows:

- Office of the Chief Surgeon
- Office of the Attending Surgeon
- Medical Supply Depot
- Base Hospital, including
  - Convalescent Hospital
  - Field Hospital No. 13
  - Evacuation Ambulance Co. 26
  - Hospital Train 70.
- Provisional Sanitary Train, 2d Brigade, A.F.G., including
  - Field Hospital 6
  - Field Hospital 8
  - Ambulance Co. 6
- Port Hospital, Antwerp
- Office of the Port Surgeon, Antwerp.
- Polish Typhus Relief Expedition (attached).
- Medical Personnel with combat units.

Polish Typhus Relief Expedition. In August 1919 this expedition had been sent to Poland to organize and conduct a campaign against typhus fever. At first its personnel comprised about 25 officers and 500 enlisted men, but in November 1919 its strength was reduced to 15 officers and 60 enlisted men.

In October 1919 the expedition was placed under the control of the A.F.G. and it continued under such control until it was withdrawn to the United States in November, 1920.

HOSPITALIZATION

After the formation of the A.F.G. a new hospitalization policy was adopted. Instead of evacuating to the United States every patient whose early return to duty was not expected, all except those whose physical condition warranted discharge from the service or required protracted observation and treatment were retained and treated in Germany. A base hospital was established, and there the sick from the various organizations were sent. The establishment of the Base Hospital (later known as the Station Hospital) assured a proper segregation of contagious diseases and provided the best modern diagnostic methods and treatment by specialists.

Hospital Train 70, having been transferred to the A.F.G., was used for the evacuation of the sick from Coblenz to Antwerp whence transports departed for the United States.

In the beginning the sick of the Base Port at Antwerp were treated in Belgian military hospitals, but this arrangement proved unsatisfactory, and in April 1920 a small port hospital was opened there. It was a well equipped hospital with a capacity for 60 beds, and it continued in operation until June 1921. Thereafter those too sick to travel were treated in the Belgian hospital, others requiring hospital treatment being sent to the Station Hospital in Coblenz.

The sick of the Polish Typhus Relief Expedition were treated in an American Red Cross Hospital in Warsaw.

Station Hosptial. Originally called the Base Hospital, it commenced to function with a staff of 59 medical officers, 54 nurses and 359 enlisted men. The bed capacity was originally 460, later being increased to 760. In 1920, in accordance with War Department orders, the name was changed to Station Hospital. In part the hospital occupied a group of buildings which, before the occupation, had composed the German lazarette (military hospital) of the garrison of Coblenz, located in the western section of the city.

Convalescent Hospital. In the latter part of 1919 the venereal rate among the American troops increased very rapidly, and it was found that the capacity of the Base Hospital was insufficient. A special hospital, which afforded quarantine facilities, was therefore established in a large and well equipped school building on Oberwerth, an island in the Rhine opposite the southern portion of Coblenz. This hospital was called the Convalescent Hospital, and in the beginning was administered, under the supervision of the commanding officer of the Base Hospital, by a line officer, the necessary medical attention being afforded by a staff of officers and men of the Medical Department. This system was soon modified, and the Convalescent Hospital was administered as a medical unit, forming an annex to the Base Hospital, two commissioned and a few non-commissioned officers of the line being retained, however. The patients were organized into a provisional infantry battalion, commanded by a captain of infantry detailed for this purpose, the company officers being officer patients. Thus, except those patients who were physically unable to participate in drills, all patients underwent a course of garrison training while in the Convalescent Hospital.

Until November 1920, all venereal patients in this hospital

were carried as on detached service from their units and as such received full pay. Afterwards they were regarded as venereal patients who were sick in hospital, and therefore, under the existing regulations, were deprived of their pay until they were returned to duty.

SANITATION.

Water Supply. This was obtained in Coblenz from five shallow wells located on Oberwerth island, the water coming from the Rhine by seepage. Gas-forming bacilli were a regular content of the Coblenz water supply, and it was necessary at all times to apply chlorination at the source. In this manner the water was kept potable until the floods in December 1919 and January 1920 put all pumps out of commission. A water point was established on the banks of the Mosel river; the water was chlorinated there and again in the Lyster bags kept by all troop units. No water-borne diseases developed among the troops during these floods.

There was a prolonged drouth during the summer of 1921, and the water supply of the American area became very scant. In June the situation at Mayen was acute, and lack of sufficient bathing facilities caused much inconvenience to the troops forming the garrison there. The water was chlorinated at the outlet of the main reservoir, and no water-borne diseases developed. As a matter of routine, all water used by the American troops was examined each week.

Food Supply. The bread furnished the troops has always been of superior quality. The refrigerated meat, brought from the United States, was usually very good. Canned goods and fruit, as well as dried groceries, were brought from the United States, while potatoes were purchased in the German market; the quality of all these was satisfactory. A limited amount of milk for consumption by the sick and young children was produced at a small dairy farm near Coblenz, operated by the Quartermaster. The milk was pasteurized and its standard was high. At the end of 1921 the dairy possessed 20 cows and had an average daily output of 40 gallons.

Disposal of Waste. Throughout the American area water closets were available in all barracks and in most billets. Many were primitive, with crude and archaic plumbing fixtures, but their purpose was adequately served. Inorganic matter was deposited on a carefully tended dump, while manure and kitchen garbage was sold to Germans who promptly removed them.

Insects. Flies, mosquitoes and bed-bugs were few and did not constitute a serious sanitary problem. Pediculi and itch mite were common, however, the latter being responsible for numerous cases of scabies.

DISEASES.

The general health of the American troops was excellent, although contagious diseases were not infrequent among the recruits received from the United States in late 1919 and early 1920. The control of venereal disease was always difficult, while diphtheria, scarlet fever and measles was ever present among the civil population.

Measles. This disease put in its first appearance in the autumn of 1919, and an epidemic occurred during the early months of 1920. The majority of the patients were members of the 5th

Infantry, which had come from the United States several months previously. During the epidemic, which ended about March 1st, 1920, 22 cases of pneumonia developed, resulting in 15 deaths. After that date measles was not a factor in the disease rate, only 5 cases occurring in 1921.

Influenza. Always present among the troops, at several times influenza appeared in epidemic form. In late 1919 and early 1920, an epidemic developed throughout the A.F.G., but most of the cases were of mild character. In March 1921 a local epidemic occurred in a battalion of the 8th Infantry; 64 patients entered hospital, 4 developing pneumonia from which 2 deaths resulted.

In December 1921 a general epidemic of a moderately severe type occurred. This epidemic had been preceded by an outbreak among the civil population, the disease spreading rapidly throughout the Rhineland and later over the entire continent of Europe. The epidemic among the American troops lasted for several weeks and resulted in 5 deaths. At its peak, on December 21, 1921, there were 212 respiratory cases in the Station Hospital. Respiratory conditions in December 1921 accounted for 694 admissions to hospital or quarters; of these 296 were influenza cases.

Pneumonia. 237 cases of primary and secondary pneumonias occurred during 1920, resulting in 36% of the total deaths of the year and 64% of deaths caused by disease. In 1921, of the 194 cases 59 occurred in December. During this year the pneumonias were responsible for 75% of deaths from disease and 37% of the total deaths.

Cerebro-spinal Meningitis. In 1919 there were only a few

cases of this disease in the A.F.G. There were but 4 in 1920, while none occurred in 1921.

<u>Diphtheria</u>. Some apprehension was caused by diphtheria in the autumn of 1919, a total of 56 cases occurring. It did not appear in epidemic form in 1920, and the cases that developed were of a mild type. 44 cases were treated in 1921, and of these 11 occurred in February.

<u>Scarlet Fever</u>. While this disease was always present in the A.F.G. an epidemic never developed.

<u>Mumps</u>. 203 cases appeared in 1920 and 224 in 1921. Of the latter 205 cases developed during the period from January to April, most of them occurring in the 5th Infantry.

<u>Tuberculosis</u>. During 1920 there were 57 cases diagnosed as tuberculosis. 4 died, and the remainder were transferred to the United States. 32 cases were similarly diagnosed during 1921, but no deaths resulted.

<u>Scabies</u>. Always very prevalent among the troops, scabies became particularly so during the fall and winter months. Severe cases were admitted to hospital, routine treatment by baths and sulphur ointment being administered.

<u>Vincent's Infection of Gums</u>. Trench mouth, as this disease is more commonly known, was responsible for many admissions to hospital in 1920, but in 1921 there was a marked improvement.

<u>Typhoid Fever</u>. No cases of typhoid fever were reported among

among the troops in 1920, but 2 developed in 1921.

<u>Venereal Disease.</u> The constant endeavors to solve this important problem were rewarded by a gradual decline of the venereal rate; at the end of 1921 the average yearly rate per 1000 was about 100. A graphic chart at the end of this chapter shows the progress that was made in the fight against venereal disease.

There were many factors that contributed to the high venereal rate that prevailed in the A.F.G. The low exchange value of the German mark resulted in the receipt by the American soldier of more pay than that of many German officials; this fact and the spendthrift ways of the average American soldier caused prostitutes from far and wide to come to Coblenz, while the open handedness of the "Yank" proved an open sesame to the favors of German girls to whom good food and clothes were worth more than their personal virtue. The illicit sale of cognac to soldiers resulted in lessened resistance to sexual desire and to infection. The venereal rate among the prostitutes was always high, while 76% of the soldiers who developed venereal disease had failed to avail themselves of prophylactic treatment despite thorough instructions and stringent orders regarding its use.

Various measures to reduce the number of exposures to venereal disease were put into effect. Women soliciting sexual intercourse and those without visible means of support were arrested and tried before a vagrancy court; those convicted were examined for venereal disease and if found infected were kept in a German hospital at Bendorf until cured. The Y.M.C.A., Salvation Army, etc., furnished a large and varied number of wholesome amusements, such as vaudeville, moving pictures, excursions and athletic competitions.

In addition to strict orders requiring them to take prophylactic treatment after exposure, every opportunity to take advantage of this preventive measure was afforded the soldiers. Soldiers were given individual packets whenever they went on pass, or at any time upon their own application; it was understood that this did not excuse a recipient from taking prophylaxis at a "prophylaxis station", as it was given only as an additional safeguard, for immediate use after exposure. Prophylaxis stations were conveniently located and thoroughly equipped for the proper application of preventive measures by a trained personnel. It was obligatory to administer prophylaxis to soldiers who returned to barracks in a state of intoxication, whether or not exposure was admitted. Prior to going on furlough soldiers were required to report to a prophylaxis station for instruction, and upon their return to report to the same place for inspection. Those who developed venereal disease and who had failed to take the required prophylaxis were tried by court-martial. Whenever the venereal rate of a unit was well above the average of the A.F.G., for any one week, a letter was written to its commander by the Adjutant, A.F.G.; this letter informed him as to the venereal rate of his command, directed his attention to the apparent reflection upon his administrative and disciplinary measures, and required that he report the cause of the high rate and the steps taken by him to remedy that condition.

In addition to all the measures outlined above, the Salvation Army gave silver cups to organizations having, among other best records, the lowest venereal rate during specified periods. Despite everything, however, the venereal rate of the A.F.G., although it improved consistently during 1920 and 1921, remained above the rate of the Army as a whole.[1]

1. See Appendix 1 to this chapter

OFFICE OF THE ATTENDING SURGEON.

This office was centrally and conveniently located and was sufficiently well equipped to provide for the medical and dental needs of the Army community in Coblenz. In addition to attending the officers and men (and their respective families) who were attached to the American Headquarters, the Attending Surgeon's Office had a large clientele of civilians attached to the A.F.G., of members of the Inter-Allied Rhineland High Commission and of members of other armies who were on duty in Coblenz. In fact, of the total cases treated only 46% were American military personnel.

The medical officers on duty in this office called professionally upon these sick in their homes and gave minor treatments at the office. In addition to their ordinary duties, they made the required physical examinations of the officers and men of the Headquarters, A.F.G.

For the greater part of the time the personnel consisted of 3 medical officers, 3 dental officers, 2-3 nurses and some 17 enlisted men. The latter included dental assistants, clerks, laboratory technicians, pharmacists and orderlies.

MEDICAL SUPPLIES.

The stock of medical supplies left by the Third Army was sufficient, with few exceptions, for the needs of the A.F.G., and comparatively little procurement was necessary. In the case of many items there was a large surplus stock, for the disposition of which considerable effort was made from time to time, but for a number of articles, especially field equipment, there was no market in Europe.

The Medical Supply Depot was located in the Proviant Magazin, a stone warehouse formerly used by the German army. A sample room, showing all types of articles on hand, was installed for the instruction of the enlisted personnel and of visiting medical and line officers. A complete field hospital, including both medical and quartermaster material, and one with medical equipment only were kept packed and assembled, ready for immediate use.

FIELD UNITS.

The provisional sanitary train which arrived in November 1919 with the 2nd Brigade, A.F.G., was assembled a year later in Neuwied. Here it came under the direct control of the Headquarters in Coblenz and was designated as the Provisional Medical Battalion.

In addition to the prescribed training, instruction in infantry drill was given. At Neuwied there was established a medical instruction center for the training of recruits of the Medical Department in the duties of sanitary troops in the field.

In 1920 and 1921 this unit participated in the A.F.G. maneuvers and held its own maneuvers as well.

DENTAL SERVICE.

Permanent base equipment was provided at the office of the Attending Surgeon and at the Station Hospital.

In addition to American military personnel and families, the privileges of the dental service were extended to members of the High Commission, to welfare workers and allied soldiers stationed in Coblenz, and to non-German civilian employees of the A.F.G.

VETERINARY SERVICE.

Until September 1920 the Veterinary Service operated independently of the Medical Department and under G-4. It was then placed under the direct supervision of the Surgeon, A.F.G., and afterwards continued as a branch of his office.

This service performed its ordinary duties, including the supervision of animal hygiene, care and treatment of sick animals, and the inspection of meats, dairy products and forage.

In May and June 1920 there was an epidemic of foot and mouth disease among the cattle and hogs at the Quartermaster farm. The infection was of mild character, and the losses were small. During 1921 there was no serious outbreak of communicable disease.

OFFICERS.

Major (formerly Colonel) E. H. Bruns, Medical Corps, served as Chief Surgeon until September 1920 when he was relieved by Colonel F. R. Keefer, M.C., who continued in that Capacity throughout 1921.

Major (formerly Colonel) R. W. Kerr, M.C., commanded the Station Hospital until September 1920. He was replaced by Lieutenant Colonel R. B. Miller, M. C.

Major (formerly Lieutenant Colonel) H. P. Carter, M. C., commanded the Convalescent Hospital from its organization until it was disbanded in August 1921.

Major (then Lieutenant Colonel) W. B. Borden, M. C., served as Medical Supply Officer until June 1920, when he was relieved by Major N. L. McDiarmid, M.C.

The following officers served at different times as Attending

Surgeon in Coblenz;

Major M. F. Welch, M. C.
Captain M. S. Weaver, M. C.
Captain F. C. Tyng, M. C.
Major W. R. Dear, M. C.

Major O. G. Skelton, D. C., was Attending Dental Surgeon in Coblenz until December 1920, when he was replaced by Lieutenant Colonel H. O. Scott, D. C.

During the existence of The Provisional Medical Battalion (November 1919 to February 1922) it was commanded by Major (formerly Lieutenant Colonel) G. L. McKinney, M. C.

APPENDIX 1

It is an interesting fact that the measures to combat venereal disease which were used by the French in the time of Napoleon I were very similar to the actions taken by the A.F.G. over 100 years later.

Venereal disease in and around Coblenz was a serious problem during Napoleonic days, as it was during the American occupation. In 1811 the Prefect of Coblenz issued an order relative to the control of prostitution which in 1919-1921 was still a city ordinance.

A circular was issued to Chiefs of police to "Awaken their sense of responsibility and their consciences to the duty of a more careful watching of debauched women and the plying of their trade with the military personnel," and "to prevent the distressing results, which were sure to obtain, when these women were infected with venereal diseases."

Provisions were made for the compulsory treatment of infected prostitutes in a Coblenz hospital, and regulations, of which the following is a resume, were published:

> All bawdy houses should be under complete control, and would not exist without the payment of certain prescribed taxes and the complete supervision of the local police.
>
> That all proprietors of such bawdy houses would notify the police commissioner of the inmates of said house, by name and in writing.
>
> No cabarets would be permitted to be maintained in these houses and no lodging of strangers nor any other specific trade could be carried on within the confines of a said registered house.
>
> All inmates of bawdy-houses would be carried on the public registers, with full information relative to their arrival in Coblenz, beginning of such trade, birthplace, last residence etc., with full information relative to parentage, health conditions, etc.
>
> The return to a life of "virtue" would in no way be

hindered by the civic authorities. A girl who demanded her liberty from a said house would immediately be released by order of the police commissioner and, if still a minor, be reclaimed by her parents or their family.

Return to the family could even be made by an adult girl, under special circumstances.

No prostitute would appear on the streets during the hours of daylight and no calling from windows or silent signalling, either by motions or signs, would be permitted.

It was absolutely forbidden for them to present themselves before or about a military confine, to be present at any troop exercises, or near schools or other establishments where youth would be congregated. No more than two prostitutes would be allowed to walk or talk together in the public streets, and no conduct that was contrary to public morals or chastity would be countenanced.

A woman plying prostitution as a trade would be obliged to notify the officials immediately upon any signs of illness, so that she could be segregated from the other inmates and treated and hospitalized until actually cured.

A special surgeon (named by official order) was assigned to the inspection of these women, which was to be held weekly.

Those discovered by the inspecting physician to be infected would be placed under arrest by a state official (unless they presented themselves voluntarily for treatment) and held until the formation of an asylum for such, which was to be called the "Salpatriere."

When women were discovered without personnally presenting themselves for treatment, they would after hospitalization, be taken by the police authorities to their birth places or legal residence under orders of expulsion.

The prices for the maintenance of such bawdy-houses would be from 200 to 500 francs yearly. Each prostitute would pay 10 francs each half-year for her health certificate and the sum of 75 centimes for each inspection, to be placed in a special fund of the Police Commission for use in the treatment of women venereally infected or for the transportation of women to be expelled from the area.

The result was not what was anticipated. Country women came to town to be registered as prostitutes. No girls asked to be returned to "lives of good morals." However, after many deportations and the hospital treatment of many women, the venereal rate displayed a marked decrease. A report by the Coblenz Hospital Administration, under date of September 23, 1811, indicated the achievement of such good results that a continuance of the foregoing measures was requested.

The above information, including the resume of the 1811 regulations, was secured from French sources in 1921.

COMMUNICABLE DISEASES 1919 - 1920 - 1921.

DISEASES	1919 Cases	1919 Rate	1920 Cases	1920 Rate	1921 Cases	1921 Rate	YEAR
CHANCROID NEW	990	8.5	772	47.7	384	25.5	1919 1920 1921
CHICKEN POX	41	.3	27	1.6	12	.8	1919 1920 1921
DIPHTHERIA	720	6.0	78	4.8	44	3.1	1919 1920 1921
GONORRHEA NEW	1913	16.4	1291	74.4	789	53.3	1919 1920 1921
INFLUENZA	2769	23.8	799	49.4	597	43.0	1919 1920 1921
MEASLES	178	1.5	118	7.3	5	.3	1919 1920 1921
MUMPS	Not reported		203	12.5	224	17.3	1919 1920 1921
PNEUMONIA	2867	24.5	237	14.6	194	13.9	1919 1920 1921
SCABIES	Not reported		384	23.7	234	16.8	1919 1920 1921
SCARLET FEVER	88	.7	54	3.3	13	.9	1919 1920 1921
SYPHILIS NEW	256	2.2	294	18.1	290	20.9	1919 1920 1921
TUBERCULOSIS	Not reported		60	3.7	72	2.3	1919 1920 1921

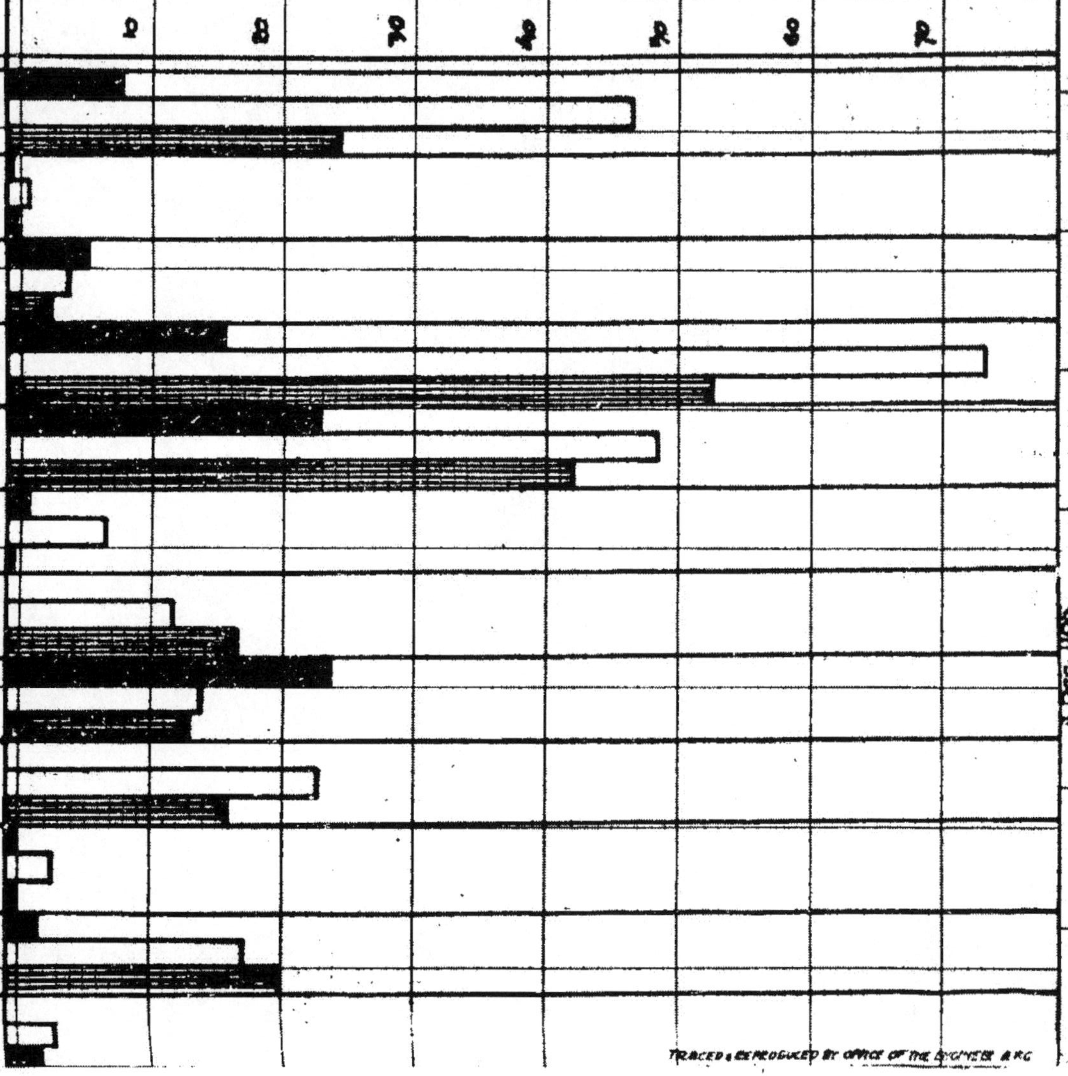

TRACED & REPRODUCED BY OFFICE OF THE ENGINEER A.R.C.

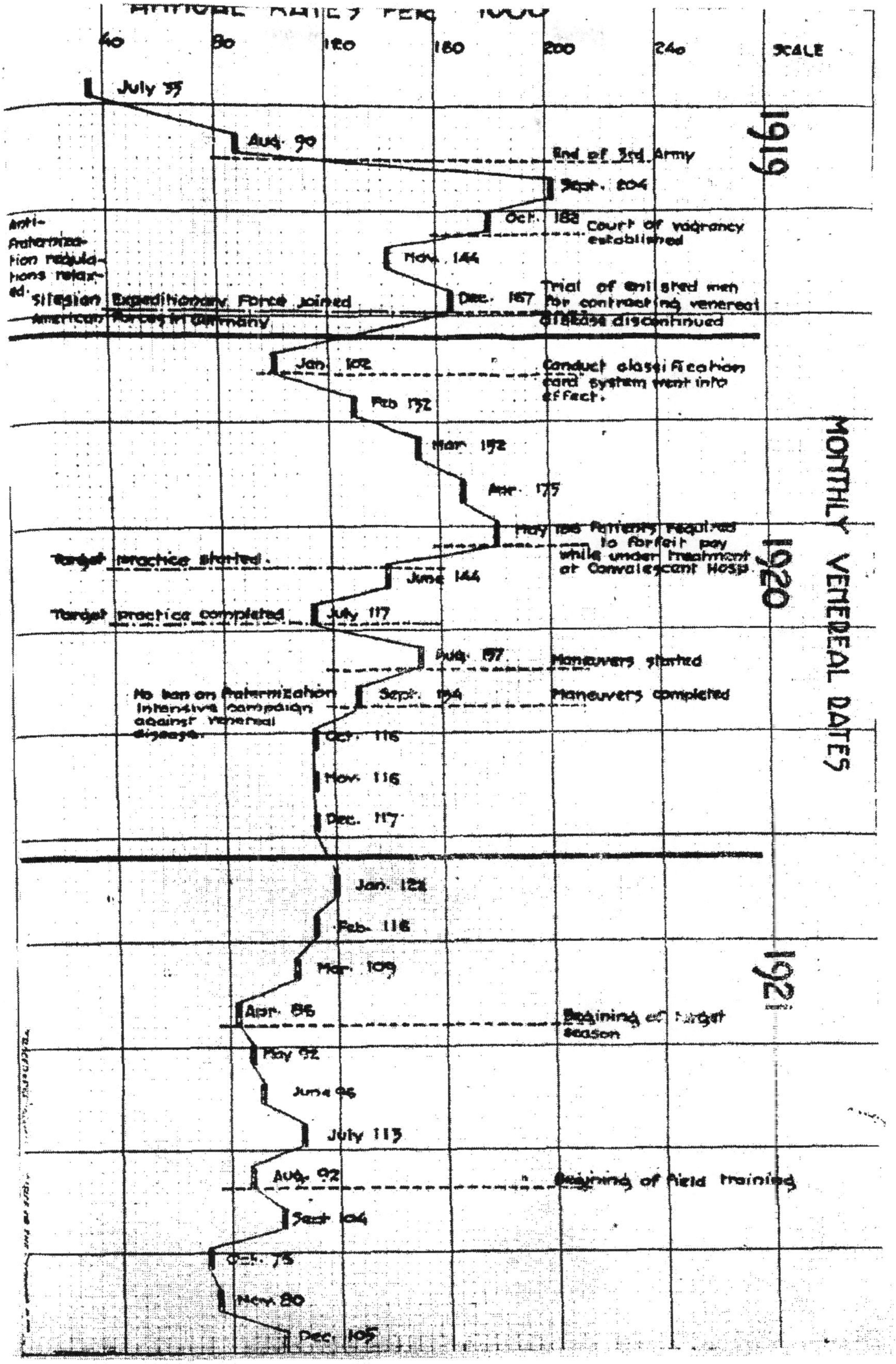

SCALE
1919
1920
1921
MONTHLY VENEREAL RATES
July 55
Aug. 90
End of 3rd Army
Sept. 204
Oct. 182
Court of vagrancy established
Nov. 144
Anti-fraternization regulations relaxed.
Dec. 167
Trial of enlisted men for contracting venereal disease discontinued
Silesian Expeditionary Force joined American Forces in Germany
Jan. 102
Conduct classification card system went into effect.
Feb 132
Mar. 152
Apr. 175
May 186 Patients required to forfeit pay while under treatment at Convalescent Hosp.
Target practice started.
June 144
Target practice completed
July 117
Aug. 157
Maneuvers started
No ban on fraternization
Intensive campaign against venereal disease.
Sept. 134
Maneuvers completed
Oct. 116
Nov. 116
Dec. 117
Jan. 121
Feb. 118
Mar. 109
Apr. 86
Beginning of target season
May 92
June 96
July 113
Aug. 92
Beginning of field training
Sept 104
Oct. 75
Nov. 80
Dec. 105

CHAPTER NO. 27

POSTAL EXPRESS SERVICE.

In the A.E.F.; In the A.F.G; Activities; Officers.

POSTAL EXPRESS SERVICE.

IN THE A.E.F.

During the World War and for some time after the Armistice the mail of the A.E.F. was handled by the Postal Express Service. Originally the mail service had been conducted by civilian personnel of the United States Post Office Department, but as the results were unsatisfactory G.H.Q. ordered that the Postal Express Service be organized. Some of the civilians were returned to the United States, while others were commissioned or drafted into the military service. The Postal Express Service was a military organization and continued so until the official dissolution of the A.E.F. The Postal Express Service operated in occupied Germany as well as in France, until the return of the Third Army units to the United States caused the great majority of the P.E.S. personnel to be sent back for discharge.

IN THE A.F.G.

A few of the experienced personnel were discharged in Europe and employed for post office work in occupied Germany, their salaries as civilian employees being paid from Quartermaster funds. When the American Forces in France were disbanded in January 1920, the Central Post Office at Paris, together with a number of civilian employees, was transferred to Coblenz. The civilians continued to be employed in postal work, but in June 1920 it became necessary, because of new regulations, to discontinue their payment from Quartermaster funds. The Post Office Department was then requested to pay the salaries of these civilian employees, which it consented to do. Instructions from Washington required that on January 31, 1922, all but one of the civilian employees should be discharged. It was then planned that after that date the mail would be handled by enlisted men exclusively, while the remaining civilian employee would perform the duties pertaining to registered mail, money orders and the sale of stamps.

In August 1919 there remained in the A.F.G. but two post offices - A.P.O. 927 at Coblenz and A.P.O. 944 at Antwerp. As so large a percentage of the experienced personnel had been withdrawn, a reorganization became necessary. A Chief of the Postal Express Service was appointed, and two Postal Officers were detailed as his assistants to conduct the post offices at Coblenz and Antwerp. Upon the arrival of the Provisional Infantry Brigade, later designated as the 2d Brigade, A.F.G., it was deemed advisable to appoint a Postal Officer for that brigade, in order that its mail service might be efficiently conducted when the brigade took station in Upper Silesia. When it became evident that the 2d Brigade would not go to Upper Silesia, its Postal Officer was relieved, and thereafter its mail service was supervised by the Chief of the Postal Express Service.

The enlisted men of the Postal Express Service in the A.F.G. were inexperienced for the most part. A school for mail clerks and mail orderlies was established and books of instruction were drawn up and printed. Organization commanders were directed to take an active interest in the solution of the problem of prompt mail distribution. A complete card index, showing the organization and station of every member of the A.F.G., was made and kept up to date. Through its use much mail that had accumulated was dispatched, after which 95% of the current mail received could be delivered or forwarded without delay.

After the disbanding of the American Forces in France and the consequent discontinuance of A.P.O. 702 (Paris), the latter was re-established by the Graves Registration Service, by which it was operated. There was mutual liaison and co-operation between A.P.O. 702 and A.P.O. 927 in regard to the handling of mail between the two. Official mail of the A.F.G. for points in France was sent to A.P.O. 702 whence it was franked by the French Postal Department.

Activities.

Up to January 1920 the Postal Express Service had maintained a courier service between Coblenz and Paris. Its discontinuance having become necessary, arrangements to use the French diplomatic courier service were made with the French postal authorities. Except on Sundays and holidays, there was a daily courier, each way, between Paris and the French Department of the Inter-Allied Rhineland High Commission at Coblenz. These couriers carried official American mail, but only four pounds of such mail could be carried by any one courier.

Early in January 1920, a daily courier service, each way, between Coblenz and Antwerp was inaugurated. These couriers carried important official dispatches for way stations, such as Andernach, Cologne, Aix-la-Chapell, etc., and official and personal mail in bulk between Coblenz and Antwerp. They also carried mail for State Department officials and military attaches at Berlin, Brussels and other points, delivery being effected through the intermediary agencies of the R.T.O.'s[1] at Cologne,

[1] Railway Transportation Officers - See Volume II, Chapters 14 and 22.

Brussels, etc.

Soon after the dissolution of the Third Army on overseas courier service between Coblenz and Washington was established, using officer couriers and enlisted assistants. Representatives of the Postal Express Service met these couriers upon their arrival in Antwerp and made all arrangements for the transfer, transportation and billeting of the couriers and their assistants. Similar arrangements were made for departing couriers. The service was discontinued in October 1921, after which all mail formerly carried by the overseas couriers was sent and received by registered mail. No disadvantage resulted; there was no delay in the handling of the mail and no losses occurred.

After the autumn of 1919 the comparative infrequency of the arrival and departure of Army transports made advisable the provision of other means of transmission of overseas mail. A request for permission to use commercial liners was disapproved by the War Department, because the necessary funds were not available. Negotiations with the Belgian postal authorities were then instituted, and they agreed to carry A.F.G. mail to the United States free of charge. This arrangement went into effect in March 1920. Between April 1 and 18, 1920, the service was suspended by the War Department, but thereafter was continued without interruption. Somewhat later the United States Post Office Department made arrangements whereby Army transports would carry mail for Belgium, Luxemburg and Belgian possessions, while ships of the Red Star Line would carry Army mail between Antwerp and New York. This arrangement went into effect in May 1920, and greatly facilitated the rapid and frequent transmission of mail between Coblenz and the United States.

Every autumn the Treasury Department appointed the Postal Officer, A.F.G., a collector of customs. As such his duties were to collect the customs duties on articles sent as Christmas presents to the United States by member of the A.F.G. If the cost of the contents of a Christmas package were certified by an officer to be not over one dollar, no duty was imposed; if the cost exceeded one dollar, duty at the prescribed rate was charged. Reports were rendered to the Treasury Department, and no irregularities were reported by that agency. This arrangement proved a great convenience, as it obviated delays in delivery by customs examinations in the United States.

Relations with the German postal authorities were generally satisfactory. A close liaison was maintained and necessary adjustments were made from time to time. A good deal of mail to members of the A.F.G. bore neither the "A.P.O. 927" nor the proper military address, and was sent by the German authorities to Hamburg in the international mail; through the liaison with the local postal officials this tangle was unravelled. Another matter was the delivery of international money orders and money remittance letters to members of the A.F.G., of which a great number came by the German post. It was arranged that they should be delivered in bulk to A.P.O. 927, by which they were delivered to the addresses, thus avoiding certain delay and possible losses.

Official American mail for points in Germany was franked by the German postal authorities; for mail going to other countries the necessary stamps were requisitioned from the German Government and placed on the letters by A.P.O. 927.

OFFICERS.

The Postal Express Service operated under the supervision and control of G-1.

Major W. A. Kenyon, P.E.S., was Chief of the Postal Express Service of the Third Army and, for a time, of the A.F.G. His duties were taken over later by Captain R. P. Bell, Infantry, who was designated as Postal Officer, A.F.G. This officer was also in charge of A.P.O. 927.

Captain W. E. Shipp, Cavalry, was Postal Officer of A.P.O. 944 at Antwerp.

CHAPTER NO. 28

MISCELLANEOUS POLICIES.

Commercial Transactions by Members of A.F.G.; Service Clubs; Employment of Civilians; Control and Conduct in Billets; Operation of American Military Police Outside American Area.

MISCELLANEOUS POLICIES

COMMERCIAL TRANSACTIONS BY MEMBERS OF A.F.G.

It came to light early in 1921 that through the medium of their German wives, several soldiers were engaging in business in the American area. As a corrective measure, General Orders No. 27 were issued on February 14, 1921 -

> Individual Commercial Transactions Prohibited.
>
> 1. It has come to the notice of these Headquarters that some members of the American Forces in Germany are engaging in business ventures in the Occupied Area.
>
> 2. Such actions are contrary to the provisions of Paragraph 338, "The Rules of Land Warfare," 1914, which is published below for the information and guidance of all concerned:
>
> "Chapter IX"
> Treatment of Enemy Property
>
> > 338. Private gain by officers and soldiers prohibited: Neither officers nor soldiers are allowed to make use of their position or power in the hostile country for private gain, not even for commercial transactions otherwise legitimate. Offenses to the contrary committed by commissioned officers will be punished with cashiering or such other punishment as the nature of the offense may require; if by soldiers, they shall be punished according to the nature of the offense."
>
> 3. These orders apply not only to officers and enlisted men of the United States Army, but to all other grades and to the civilian personnel employed by, attached to or accompanying the United States Army in the occupation of Germany.
>
> 4. It is further directed that immediate action be taken to terminate any existing business ventures or commercial transactions that are in conflict with the letter or spirit of this order.

The provisions of Paragraph 338 of "The Rules of Land Warfare" were originally published in General Orders No. 100 (Article 46), War Department, 1863.

SERVICE CLUBS.

In addition to the Officers' Club, there were three service clubs in Coblenz:

Field Clerks' Club
Staff Non-Commissioned Officers Club
(for Grades 1 to 3 inclusive)
Sergeants' Club
(for Grades 1 to 4 inclusive)

In both Andernach and Mayen there were clubs for officers and for non-commissioned officers.

The policies governing the clubs for field clerks and non-commissioned officers were prescribed in the following rules which were officially furnished on November 21, 1920, to all unit commanders for publication to those concerned:

Rules for Service Clubs

Field Clerks Club

1. Membership will be restricted to Warrant Officers, Army Field Clerks, Quartermaster Corps, of the United States or Allied Armies, and to civilians who have been honorably discharged while holding one of these grades in the United States or Allied Armies, or who are employed by the United States or Allied Armies in positions similar to those occupied by Field Clerks.

2. Male visitors will be restricted, subject to such rules as the Field Clerks Clubs may adopt to the persons mentioned in Section I, Par. 1, this memorandum.

II. Staff Non-Commissioned Officers Clubs.

1. Membership will be restricted to non-commissioned officers actually of the first three (3) grades, as set forth in General Orders #36, W.D., 19 June, 1920, of the United States or Allied Armies, and to civilians who have been honorably discharged while holding one of these grades, in the United States or Allied Armies.

2. Male visitors will be restricted, subject to such rules as the Staff Non-Commissioned Officers Clubs may adopt, to the persons mentioned in Section II, Par. 1, this memorandum.

III. Sergeants Clubs.

1. Membership will be restricted to non-commissioned officers actually of the fourth

(4th) grade or higher grades, as set forth in General Orders #36, W.D., 19th June, 1920, of the United States or Allied Armies, and to civilians who have been honorably discharged while holding this grade in the United States or Allied Armies.

2. Male visitors will be restricted, subject to such rules as the Sergeants Clubs may adopt, to the persons mentioned in Section III, Par. 1, this memorandum.

In exceptional cases, upon the unanimous approval of the interested club, applications may be made for permission for worthy American or Allied civilians to be made a member of a suitable club. These applications, setting forth fully the reasons for the request, will be submitted to the officer under whose directions the club operates for approval, and will be carried on a list in the records of each club as "Special Member."

All present members of Service Clubs, who do not conform to these requirements as to membership, will be dropped on or before 1 December, 1920.

EMPLOYMENT OF CIVILIANS.

During the latter half of 1921 and the early part of 1922, there was a somewhat indiscriminate employment of civilians, especially Germans, in organization messes and barracks. This was largely due to the reduced strength of the organizations, which made it difficult for their personnel to perform the necessary tasks. It was found advisable to issue certain restrictions upon the employment of civilians for such purposes. The first of these was contained in a letter to all organization commanders, issued over the signature of the Adjutant on July 29, 1921 -

The Commanding General directs that all female civilians employed in U. S. Army messes be discharged prior to August 15, 1921, except those employed in the Nurses' mess and the patients' mess at the Station Hospital, Coblenz, where female civilians may be employed with the approval of the Station Hospital Commander.

Further restrictions became necessary, and Circular No. 6 was issued on January 7, 1922:

Employment of Civilians.

1. It has been brought to the attention of these Headquarters that young boys of foreign nationality are being employed by organizations of the command as mess attendants, etc. In several instances it has been found that these boys have escaped from reform schools and have no identification papers whatsoever in their possession.

2. Hereafter no person will be employed: -

(a) Under eighteen years of age

(b) Without having in his possession a certificate issued by Oberstadtsecretar Rauen, Bureau 3, for the City of Coblenz and by corresponding officials for points outside of Coblenz.

3. The regulations relative to the non-employment of women remain in effect.

4. No persons unemployed or under eighteen years of age will be permitted to make a rendezvous of the compounds or barracks.

On March 16, 1922, the rules governing the employment of civilians were condensed in Circular No. 33 -

Employment of Civilians.

1. Circular Letter, subject "Female civilian employees in Army Messes to be discharged," Headquarters, American Forces in Germany, 29 July, 1921 and Circular No. 6, Headquarters American Forces in Germany, 7 January 1922, are rescinded and the following instructions are issued in their place.

2. Female civilians will not be employed in any United States Army Organization Mess, maintained in this command, except that, with the approval of the Commanding Officer Station Hospital, female civilians over forty (40) years of age may be employed in the Nurses' Mess and in the Patients' Mess at the Station Hospital, Coblenz.

3. Male civilians will not be employed in any United States Army Organization Mess, Post or Compound, situated in this command, unless they are over eighteen (18) years of age and

have in their possession a certificate issued by Oberstadtsecretaer Rauen, Bureau 3, for the City of Coblenz, or by a corresponding official for points outside of Coblenz.

4. Commanding Officers will take the necessary action to insure that no unemployed civilians and no persons under eighteen (18) years of age are permitted to loiter about barracks, posts and compounds.

5. Voluntary contributions from the enlisted men of a command donated to pay the wages of civilians authorized to be employed, as Mess Attendants, Kitchen, Stable and Barrack Police, will be taken up on organization fund accounts by proper invoice and expended by proper receipt.

6. In addition, where meals are furnished the various authorized civilian laborers by organization messes, the organization fund will be reimbursed from the amount collected by voluntary contribution to an amount equivalent to the value of the rations furnished the laborer. In garrison, the sum due will be calculated upon the current value of the garrison ration as published monthly in orders from these Headquarters. In the field, where civilian employees are fed by organization messes, the value of the field ration will be used as a basis for this calculation.

7. Organization commanders will be held responsible that the employment of civilians for kitchen and stable police, and other such duties, is not carried to such an extreme, as to prevent the training of enlisted men of their commands in these necessary duties of a soldier.

CONTROL AND CONDUCT IN BILLETS.

In December 1921 a serious dispute, involving an assault, occurred between a French employee of the High Commission and a German in whose house he was billeted. The French High Commissioner requested certain action which would have interfered with the control of billeting in the American area, always maintained by the Commanding General of the A.F.G. The Chief of Staff referred the case to the Inspector by the following memorandum:

1. The Commanding General directs that you make an investigation, report of, and recommendations

concerning the controversy between-----------,
of the French Rhineland Commission and-------,
of German house holder. All the papers in the case
are forwarded herewith.

2. The Commanding General's policy in regard to the relations between Germans and members of the American Forces in Germany is fixed in regard to the following points:

(a) These Headquarters control and decide all matters concerning billeting, and members of the Allied Commissions are subject to the American Billeting regulations.

(b) No person billeted in a house where members of German family are also billeted has any right to give any orders whatsoever to the members of said family, in regard to the use or disuse of parts of the house which are intended for joint use of the two families.

3. It is desired that this investigation be thorough, but that it be completed within the shortest possible time consistent with a thorough investigation.

OPERATION OF AMERICAN MILITARY POLICE OUTSIDE AMERICAN AREA.

On October 5, 1920, an American military policeman arrested a British subject (negro) in a cafe in Antwerp, Belgium. The incident was reported to the British Consul there and was the subject of investigation and correspondence. To prevent other international complications of this nature the following instructions were issued to the Provost Marshal of the A.F.G.

1st. American Military Policemen and members of the Division of Criminal Investigation, serving outside of the American Occupied Area, will not arrest, molest or question with a view to arrest any person, who is not in the uniform of the United States Army or who, if in civilian clothes, is not positively known by them to be a member of the United States Army, except when they actually detect a person forcibly entering upon or being unlawfully present upon property rented, requisitioned, loaned to or owned by the United States Government.

2nd. American Military Policemen and members of the Division of Criminal Investigation, serving outside of the American Occupied Area, observing a misdemeanor or a felony committed by a person in civilian clothes, when they have reason to believe is a member of the United States Army, or is a deserter therefrom, will obtain the services of the local civilian or Military Police in order to make the actual arrest of the offender.

3rd. American Military Policemen and members of the Division of Criminal Investigation, serving outside of the American Occupied Area, are authorized to arrest and confine, upon receipt of a written order or request, signed by the proper commissioned officer or American Consul, any person belonging to the United States Navy, Marine Corps, Coast Guard or Merchant Marine Services, the action taken with its attendant circumstances being immediately reported to higher authority.

Made in the USA
Monee, IL
07 November 2026

68002973R00234